A New Vision of the Early Universe

Third Edition

Also by
Robert J. Conover

Strategic Tort
Mediation Negotiation

Journey of the Universe
Second Edition

A New Vision of the Early Universe

Third Edition

The Foundation for a New Quantum Reality

InPerspective Publications

Cover image from Pixabay.com
Illustrations of photon & neutrino by Anna Keene
Editing by Donna D. Lawson

ISBN-13: 979-8-9918554-0-2 (Hardback)
ISBN-13: 978-0-9797298-4-3 (Paperback)

Library of Congress Control Number: 2024926224

Third Edition
Printed in the United States of America
March 2025

Science/Space Sciences/Cosmology
Philosophy/Metaphysics

InPerspective Publications
PO Box 805
San Luis Obispo, CA 93406
www.inperspective.space

In memory of my father, who encouraged me to learn all that I could, and taught me the value of perspective.

And too my friends Socrates, Newton, Einstein, and many others, who by example encouraged me to pursue new perspectives.

And to my wife, Ellen, without whose patience, love, and support, this project would not have been possible.

Contents

Preface

Introduction   1

Part I
Building a Foundation

One – 1.0 **Fields & the Double Slit Experiment**   11

1.1 The Origin of Particle Fields

1.2 The Double-Slit Experiment

1.3 Experimental Predictions & Conclusions

Two – 2.0 **Choosing the Initial Conditions**   23

2.1 Initial Conditions

2.2 What is Energy?

2.3. An Overview of Penergy

Three – 3.0 **The Origin of Supermassive Blackholes**   41

3.1 Blackholes

3.2 Arguments for SMBH Creation by Mergers

3.3 Alternative Theory for SMBH Creation

3.4 Spin/Density Relationship with Penergy Medium

3.5 Evidence for SMBH Creation & Cascade Effect

Four – 4.0 **Elem. Particles & Intergalactic Medium**   65

4.1 Early Elementary Particles

4.2 The Mystery of the Missing Anti-Matter

4.3 The Intergalactic Medium

4.4 Conclusions to Part I

Part II
The Heart of the Puzzle

Five 5.0 **The Theoretical Basis for Force**   83

5.1 Force According to the Standard Model
5.2 Force According to the Tor Model

Six – 6.0 **The Early Cosmic Environment**   111

6.1 Survival by Natural Selection
6.2 Evolution Progresses in Levels
6.3 The Purpose of Entropy
6.4 Emergent Qualities
6.5 Cosmic Homogeneity

Seven – 7.0 **Configurations: Photon & Neutrino**   123

7.1 Arguments for Composite Particles
7.2 Particle Configurations
7.3 Photon Configuration
7.4 Connector/Neutrino Configurations

Eight – 8.0 **Config: Quark, Electron, Nucleus & Atom** 141

8.1 Constructing Quark Charge Values
8.2 Quark Configurations
8.3 Electron Configuration
8.4 The Nucleus and Atom
8.5 Observational Evidence for Big Bang Theory
8.6 Conclusions to Part II

Part III
Bringing the Puzzle Into Focus

Nine – 9.0 **The Reality Puzzle**   163

9.1 Finding Reality
9.2 Quantum Theory
9.3 The Reality of Quantum Theory

Ten – 10.0 **The Big Picture**   185

10.1 In Search of the Big Picture
10.2 Cosmic Perspectives
10.3 Cosmic Reality

Part IV
The Appendix of Puzzling Cosmological Issues

Appendix

#1 Origin of Mass   201
#2 Origin of Dark Matter   215
#3 Origin of Gravity   221
#4 Why Space & Time are Relative   235

Notes & References 245
Bibliography 253
Index 259
About the Author 266

The initial conditions provide the first images and mathematical edges from which to connect the rest of the puzzle. If we get those puzzle pieces wrong, every connection that follows becomes a conceptual challenge, requiring far too many hypotheticals and creative mathematics, which pretty much describes our current vision of the early universe.

Preface

Cosmic reality comes to us in philosophical and scientific bits and pieces, one accomplishment at a time, like filling in a cosmic jigsaw puzzle. We mortal puzzle assemblers have long sought to discover the bona fide puzzle pieces and fit them together. Philosophers over two millennia have portrayed puzzle images and created stories to link them together. Only within the last five hundred years have we discovered a scientific approach to puzzle assembly. Puzzle pieces have a mathematical edge to them, and with the correct mathematics we can fit the pieces together with an exact fit.

Cosmic puzzle assembly has evolved into either connecting adjoining pieces by painting a reasonable picture, or by examining the mathematical edges to each piece and fabricating new tight-fitting pieces. The problem with the former assembly technique is that despite the pretty picture produced, the edges of painted pieces don't always fit together smoothly. The problem with the later assembly technique is that our newly manufactured pieces fit together precisely, but we can't be sure the resulting image is an accurate reflection of reality. And, both techniques occasionally require hypotheticals to mend, bridge, or explain bad fits or images.

To see the universe as it was, is, and will be, we need the puzzle pieces to fit together with mathematical precision while giving us a reasonable picture of cosmic reality. Philosophers want to paint an image and hope it fits, and physicists want to find a fit and hope it portrays a tenable image. The latter approach has had some wonderful successes, but lately the growing toleration and dependence on hypotheticals has blurred the picture, giving us questionable interpretations of reality. Modern philosophers have not fared well either. The puzzle has become so complex that trying to paint an image that encompasses the many theories believed to be valid puzzle segments has proven nearly impossible.

The core segments of the puzzle have been complete for many years. Maxwell and Einstein's work on fields are beautiful segments that fit together tightly. Quantum Electrodynamics (QED) and related segments explaining the photon's interaction with matter required some mathematical fudging but otherwise presents a beautiful image. Einstein's work on Special Relativity is tight. The mathematics of many other segments is very good, but all too often they require improbable hypotheticals making the image a bit uncertain. The cosmic puzzle is good but remains incomplete. Despite the efforts of many scientists and mathematicians, little real progress has been made in the past forty years. Some would say we have discovered more loose puzzle pieces than we have put together. The segments portraying the early universe in particular are fraught with unanswered questions, conflicts, and mysteries, requiring many hypotheticals to hold them together. Something is clearly missing from that part of the puzzle. We've tweaked our current vision to where it is satisfactory but well short of complete or compelling. Perhaps we need a whole new vision of the early universe.

To ensure the puzzle pieces fit together tightly, mathematics is ultimately necessary, but it can get in the way when developing a new model. Attaching a new model mathematically to the existing semi-working model is hazardous because one does not always know why or where in the mathematical scheme the old model went wrong.

Additionally, assemblers fall in love with questionable segments they are hesitant to get rid of, skewing progress. Believing current equations and images are too beautiful to be wrong is a notorious fallacy in puzzle assembly.

Again, both puzzle assembly methods have their faults: philosophers provide pictures without mathematical precision and physicists provide mathematical precision without an assurance of reality. So, who do we trust to overcome the current quagmire of unanswered questions, conflicts, and mysteries plaguing our early puzzle picture?

We can't trust either method completely. Recent mathematical successes have drawn us away from the philosophical approach, but our failure to progress in assembling a more cohesive early puzzle suggests we need to look at any approach that shows promise of moving us forward. As Cosmologist Carolyn Devereux reminds us, "the basic scientific process is drawing rational conclusions from observations in an objective way.[1] We should be able to employ that process using either puzzle assembly method.

While reading Michio Kaku's, *The God Equation*[2], I realized the reason all the puzzle pieces were not fitting together was because we may be framing the cosmic picture incorrectly. The first and most important puzzle pieces are those that frame the overall image, which for our picture is what we ascribe to initial conditions. The initial conditions provide the first images and mathematical edges from which to connect the rest of the puzzle. If we get those puzzle pieces wrong, every connection that follows becomes a conceptual challenge, requiring far too many hypotheticals and creative mathematics, which pretty much describes our current vision of the early universe. Once I returned to the universe's moment of birth, resolved the initial conditions, and allowed the universe to naturally evolve, the puzzle pieces began falling into place. This is the story of that natural evolution. It smoothly transitions into the universe we see today and retains the observational evidence supporting the Big Bang Theory.

The third edition of *A New Vision of the Early Universe – The Foundation for a New Quantum Reality,* remains framed in the story of the first 380,000 years but it is not so much about evolution as it is about how the universe works. There is more focus on solving cosmological mysteries and clarifying concepts, including the foundation of quantum mechanics. It is not only a new vision of the early universe, but a new vision of quantum reality.

Hawking's conclusion is monumental. It reveals the mistaken assumptions made by Lemaître, Gamow and other architects of the BBT.

Introduction

Our initial sequence of cosmic puzzle pieces is arranged by our current cosmological model, which is commonly known as the Big Bang Theory (BBT). For the past hundred years we have tweaked and massaged the BBT into a plausible and supported hypothesis. Yet it contains many paradoxes and mysteries, and lately it seems to acquire more problems than solutions. As Physicist Stuart Clark observes, "There are a growing collection of observations that cannot be squared with today's theories."[1]

Much of the observational evidence for the BBT comes from conditions occurring 380,000 years after the big bang. Those conditions (*cosmic expansion, atom ratios, the CMB, etc.*) indeed support the BBT but that same evidence could support several other theories. Notwithstanding that observational evidence, the BBT is obviously incomplete. It leaves many unanswered questions and fails to explain such things as our observed large-scale structure, the uniformity of the CMB, dark energy, dark matter (*if it exists*), the missing anti-matter, or why the James West Space Telescope (JWST) is finding huge galaxies in existence so soon after the big bang. Many authors echo the concern of Physicist Harry Cliff, who informs us, "Something deep is clearly missing from our current telling of the cosmic story,"[2] The BBT has served us well but it is surely not the final vision of the early universe. It may be time to explore alternatives.

Apparently, the validity of the BBT depends on the successful development of Quantum Gravity. According to Physicist Brian Clegg, "Without quantum gravity, we can't say what happened at the big bang or even if it existed at all. The only certainty is that our current scientific models fail. Entirely. The same goes for singularities at the heart of blackholes, not to mention some of the more exotic concepts in cosmology...."[3.]

Another attack on the validity of the BBT comes from Stephen Hawking. *The laws of physics are not fundamental but are emergent.* That is the final conclusion of the late Astrophysicist, as relayed through his working associate, Thomas Hertog, in his recent book, *On the Origin of Time.* Hawking's conclusion is monumental. It reveals the mistaken assumptions made by Lemaître, Gamow and other architects of the BBT. From Hawking's conclusion it's evident that employing Einstein's equations at the very beginning of the universe may not have been valid since the laws of physics, as well as matter and gravity, had not yet emerged at the instant the universe began. General Relativity was certainly important to the early universe, but not at the moment of its birth. Hawking's conclusion implies the universe experienced some undetermined evolution prior to those things emerging. His conclusion will change our conception of the initial conditions and how the early universe evolved. Notwithstanding its accumulated evidentiary support, it appears the BBT may have gotten off on the wrong foot.

Personally, the most telling flaw in the BBT is that it doesn't support a clean and direct story of evolution. It posits things into existence with ambiguous causal connections to a prior related condition. Positing a hypothetical condition is sometimes necessary to explain a theory, but hypotheticals should be considered temporary placeholders in the cosmic puzzle, relied upon cautiously until that area of the puzzle can be supported by evidence. Extrapolating hypothetical upon hypothetical is a dubious exercise leaving us open to picturing a mathematically supported but otherwise non-existent reality. We have found a way to make the mathematics work but given the limited observational or experimental evidence supporting the BBT, those puzzle pieces may not reflect the real story.

The lack of evidence and the dependence on so many hypothetical conditions to support our core theory describing the early universe leaves us in desperate need of answers. We are looking to the Large Hadron Collider (LHC) for new physics; to quantum gravity for new a mathematical architecture; and to a TOE or other exotic theory, such as String Theory, Supersymmetry, Loop Quantum Gravity etc., for the big picture. While all those pursuits are worthwhile, they are starting to look like quests for miracle cures. The number of missing pieces, conflicts, and weak connections in our picture puzzle, most of which have plagued us for decades, suggest we may simply be on the wrong track.

So, where do we go from here? According to Brian Clegg, the BBT is invalid without the illusive Quantum Gravity, which implies it is not a simple fix and we may have to start over. According to Stephen Hawking, the initial assumptions used to build the BBT may have been invalid, which again suggests we should start over. The observation that the BBT does not provide a clean evolutionary sequence and requires many hypothetical conditions for support also suggests we should consider starting over. As recently observed by Physicist Harry Cliff, "The Hubble tension and the problem over lumpiness of the universe seem to be pointing strongly to the fact the standard cosmological story is ripe of rewriting."[4.] And as Physicist Sean Carroll reminds us, "Any of our theories could be wrong, no matter how much evidence we have so far accumulated for them."[5.] As observed by others, although we have no alternative to the BBT in sight, that is no reason to accept it.[6.] We need not set out for a total replacement of the

BBT, but when feasible we should examine new perspectives that might solve some of our outstanding mysteries.

A new perspective on the early universe is now possible due to the recent discovery that universal expansion is accelerating, from which emerged the concept of *Dark Energy*. Realizing the existence of a thin density, fluid-like, intergalactic medium makes all the difference, giving us a perspective the architects of the BBT did not have a century ago. Dark energy turns out to be the missing puzzle piece that links more elegantly the segments representing particles, fields, and forces, and it illuminates the foundation of quantum mechanics. In the long run, the model herein may not describe the exact evolution of the early universe, but it answers many questions and demonstrates the feasibility of a cohesive story that smoothly brings the elements of the universe naturally into existence from an evolutionary basis.

We are starting over with nothing more than a blank slate and two fundamental principles. We are writing a new story and creating a model that starts quite differently but ends in the universe we know today with many of its mysteries resolved. The observational evidence and the richness and simplicity of the model will garner additional experimental and mathematical support from those theorists seeking a new perspective.

Because the model is a substantial departure from current theory, I shall begin with an examination of the aspect that best lends itself to verification, which is the subject of *fields* and the resolution of the issue of wave-particle duality. Starting the story well into the evolution of the early universe is a bit awkward but the predicted outcome of the experiment explains a great deal and lays a good foundation for the rest of the model.

Consequently, we will start the story in the middle, at a point when the only important things to have evolved are particles and a dark-energy-type intergalactic medium (IM), both of which we know exist. Exactly how each of those evolved will be taken up in due course. Many of the points made will be in contrast to the Standard Models, and since the Standard Model of Cosmology and the Standard Model of Particle Physics borrow heavily from each other regarding the early universe, I will henceforth refer to them together simply as the Standard Model. Again, to differentiate the pure energy that started the universe from other forms of energy we may discuss, I will refer to pure energy as *Penergy*.

If you are a dedicated cosmic puzzle assembler, cosmological mystery solver, or simply someone interested in how all the puzzle pieces might fit together, you will enjoy this read.

One of the goals of this treatise is to solve some of the outstanding cosmological issues either directly or by providing food for thought for its later resolution by looking at it from a fresh perspective. Each of these thirty-four issues will be addressed and the issue checked off at the end of the applicable chapter.

Resolution of
Cosmological Issues

- [ ] Explanation of Work Energy
- [ ] Foundation of Quantum Mechanics
- [ ] Magnetic Monopole Mystery
- [ ] Missing Anti-Matter Mystery
- [ ] Neutrino Oscillation
- [ ] Origin of Cosmic Web
- [ ] Origin of Dark Energy
- [ ] Origin of Dark Matter
- [ ] Origin of EM field
- [ ] Origin of EM Waves
- [ ] Origin of Fields
- [ ] Origin of First Stars
- [ ] Origin of Force
- [ ] Origin of Gravity
- [ ] Origin of Mass
- [ ] Particle Decay
- [ ] Particle Evolution Sequence
- [ ] Particle Tunneling
- [ ] Purpose of Entropy
- [ ] SMBH Creation
- [ ] Space & Time Relativity
- [ ] The Fine-Tuning Problem
- [ ] The Flatness Problem
- [ ] The Hierarchy Problem
- [ ] The Horizon Problem
- [ ] The Magnetic Field
- [ ] The Measurement Problem
- [ ] The Reality Problem
- [ ] Virtual Particles
- [ ] Wave-Particle Duality Resolution
- [ ] What Drives Cosmic Evolution
- [ ] What is Energy
- [ ] Why Gravity so Weak
- [ ] Why Photons are Massless
- [ ] Why Three Generations

Part I

Building a
Foundation

Those probabilities are accurate using wave-like mathematics because the particle's movements, its field, and the intergalactic medium are all behaving in a wave-like manner.

Chapter One
1.0 Fields & the Double Slit Experiment

The basic hypothesis for this new model of the early universe is that the initial speck of Penergy expanded, lost density, and spun itself into supermassive blackholes (SMBHs) and particles, leaving a thin density intergalactic medium (IM). There is substantial observational evidence supporting that evolution, which will be covered in chapters three and four. For now, to examine the origin of fields we need only assume the existence of particles and a dark-energy-type intergalactic medium.

The contents of the Penergy speck was consumed in the creation of SMBHs, particles, and the IM. It is from relationships between them that everything else in our universe evolved. The first relationship to evolve was between particles and the IM, which created *fields*. This notion is presented in contrast to Quantum Field Theory's (QFT) concept of fields.

1.1 The Origin of Particle Fields

Opinions differ as to how fields are created. Do fields create particles, or do particles create fields? In QFT, fields create particles. Those ubiquitous fields were created at the big bang, exist throughout the universe, and for every particle there is a specific field. Particles are merely excitations of their ubiquitous field.

The new model sees the creation of fields differently, but there are similarities between the two theories. QFT recognizes that particles and fields are associated, but it is the fields that are ubiquitous throughout the universe and from which everything else is made.[1] In the new model, the IM is ubiquitous throughout the universe and from which everything else is made. In QFT, the ubiquitous fields are the elements that play the role of a communications network between particles.[2] In the new model, the IM is the element that plays the role of a communications network between particles. Given the evidence for the existence of Dark Energy, the presence of an IM seems very real. There is no evidence for ubiquitous fields aside from QFT equations.

QFT is a mathematical edifice that describes a reality difficult to believe. Physicist George Musser tells us physicists struggle to understand what quantum field theory is telling them about the world.[3] Chemical Physicist Michael Munowitz says that "a quantized field at first blush is a monstrosity, a mathematical absurdity that makes no sense at all."[4] Physicist Sean Carroll tells us, "The Core Theory. [*Quantum Field Theory*] is not the most elegant concoction that has ever been dreamed up in the mind of a physicist, but it's been spectacularly successful at accounting for every experiment ever performed in a laboratory here on earth."[5] Quantum math is torturous, but it works. But that success does not mean that QFT's interpretation of how that math reflects reality is correct.

QFT might be mathematically accurate, but it is equally feasible for fields to arise naturally due to the relationship between particles and the IM. As initially imagined by Michael Faraday, a particle's field could be simply a local disturbance in the intergalactic medium due to its vibrating presence. Let's examine what that might look like.

Fields according to the new model are not ubiquitous but are personal to an individual particle or group of particles making up any size body. The field can be viewed as a bubble around the particle but there is no physical bubble per se, it is only a visual image. We can't see individual particle fields, but we know enough about how they affect other particles to have a vision of their presence. There is no dispute as to particles having fields, only the source of those fields.

The argument for a particle having a local, bubble-like field is found in our understanding of *particle location*. According to Quantum Theory, the location of a particle is not defined until we measure it, and particle locations are determined by waves of probability. As scientists have observed, particles can be pushed through the same experiment over and over and come out in a different place each time.[6.] Quantum Theory ascribes this phenomenon to the particle also being a wave, which gives it an indeterminate location. Given the current belief in wave-particle duality, that explanation is reasonable but not necessarily accurate.

Scientists never see particles as waves, only as point-like particles.[7.] In other words, there is no physical evidence that a particle is also a wave. Particles are believed to be waves based on the interpretation of experimental outcomes, such as in the double-slit experiment (*discussed shortly*) and from the experience of a repeated experiment resulting in different location outcomes. These experimental outcomes seem counterintuitive and weird, fostering a wave-like interpretation, but they can be explained by visualizing a particle that is producing its own individual, local field within the IM.

A particle is not a wave. It can seem to behave like a wave because its field, created by the particle's spinning gyrations disturbing the IM, has the characteristics of a wave. As the particle corkscrews through the medium, it constantly creates a bubble of disturbed medium around itself. That relationship causes the oscillations of the particle to be imparted to the fluid-like IM. This is just what we see in the real world. Astrophysicist James Geach tells us, "... for a charged particle surrounded by an electrostatic field, any oscillation of the particle will also oscillate the field."[8.] This suggests the particle is the cause and the field is the effect.

The bubble of disturbed IM comprising the field is not fixed. It is very fluid, and its dimensions, shape, and behavior are constantly in flux. The IM is a frenzy of activity due to the volume of particles, their extended fields, related forces, gravitational waves, energy fluctuations, etc. Consequently, the bubble is constantly distorted depending on what conditions the particle is responding to as it progresses through the frenetic environment of the IM.

Again, the particle's gyrations create vibrations that change the state of the IM, creating a distinct environment that defines the bubble-like field. Within that bubble environment the particle is free to roam while at the same time seemingly confined by the bubble. While the particle is creating the bubble, the bubble is giving the particle a safe and defined place in which to reside. It is an extension of the particle, like a barrier protecting the particle - a first line of defense and communication when encountering another particle. The practical result is that the particle can be found anywhere within its bubble-like field, with its precise location determined by mathematical probability. Those probabilities are accurate using wave-like mathematics because the particle's movements, its field, and the IM are all behaving in a wave-like manner.

Given the above description of a particle's personal field and the idea that fields are without boundaries, it is understandable why quantum theory posits particles being anywhere until they are measured or detected. Fields such as the electric field may be without a boundary technically, but the density of the field falls off so quickly that its practical boundary, where it has an insignificant influence on other particles, is not far from the particle itself. Therefore, the probability of finding the particle far away from the center of its field becomes remote. Again, at any moment the particle might be found anywhere within its field, which explains why one would obtain different location-results with an identically repeated experiment.

Another instance of particle behavior suggesting it is a wave regards its apparent capacity to tunnel into places to which it theoretically should be incapable of moving. A particle shot at the edge of a wall will usually bounce back, as we would expect a tennis ball to do. But occasionally the particle can be experimentally found on the other side of the wall. This is an example of *particle tunneling*.

In quantum theory the interpretation of this example of tunneling is that the particle is a wave that occasionally *diffracts* around the edge of the wall, as waves are known to do. Its bending around the edge of the wall allows the particle to show up on the other side. This seems only possible if the particle is a wave. In the new model, it is the particle's field that is diffracting around the edge of the wall. Since the particle is seldom located at the edge of its field, it is seldom that it is within the portion of the field that is being diffracted. Occasionally, however, the particle is at the edge of its field that is being diffracted and it shows up on the other side of the wall.

Scientists have seen evidence of particles behaving like waves for some time. A major piece of evidence supporting the theory was the very reasonable interpretation of the *double-slit experiment*. Let's examine the double-slit experiment and evaluate whether an alternative explanation for its outcome is feasible.

1.2 The Double-Slit Experiment

One of the cornerstones of Quantum Theory is the idea that particles are both a point-like particle and a wave, giving their behavior a great deal of uncertainty and possibly meaning particles are not real at all. True enough, particles exhibit point-like particle attributes in scattering experiments where they ricochet off each other. And they exhibit wave-like attributes in experiments like those involving electron beams striking crystalline solids in which the electrons exhibit diffraction and interference, both attributes found in waves. But perhaps there is a plausible explanation for this apparent wave-particle duality.

The proof of the wave-particle duality comes in the way of the double-slit experiment. When a wave is washed against a thin wall having two closely placed vertical slits, the slits will cause the passing wave to diffract as it passes through. The semicircular waves emerging from the two slits will interact with each other as the wavefronts cross each other's path. The diffracted waves from the two slits interfere with each other both constructively and destructively, creating crests and troughs that show up vividly on a detection screen set just beyond the wall. This setup is known as the double-slit experiment.

When sending either photons, electrons, or even larger particles through a two-slit setup, the detection screen will show the wave-like pattern of crests and troughs. The conclusion drawn from these experiments is that the individual particles (*or their wave functions*) are passing through both slits at once, interfering with themselves, and therefore are not only acting like waves, but *are* waves. They originate as particles, and end as particles at the detection screen, but in between they act like waves.[9.]

That conclusion is understandable but may not be correct. One can also imagine the same experimental outcome with each particle passing through only one slit but its personal field passing through both slits. The personal field is split by the two slits and diffracts around their edges. The split field continues forward under its own momentum and strikes the detection screen. The wave-like appearance of particle locations on the screen would be consistent with the probability of finding the particle greater in the constructive portions and lesser in the diminutive destructive portions of its field, hence the detected interference pattern.

It is also noted that in a separate experiment if one places a detection device at each slit to determine which slit each particle went through, the interference pattern does not show up. The reason for this may be that the detection device caused the particle's field to be significantly disturbed and effectively collapsed at the entry to the slit, thus the field did not go through both slits and therefore could not diffract and produce the interference pattern.

1.2a Particle Field Collapse

Let's examine what is meant by particle field collapse in this instance. Recall that the spinning undulations of the particle disturbs the surrounding IM creating a tension in the medium that defines the particle's field. The particle will always produce a field, but that field does not have a precise boundary. In circumstances where two or more particle fields are forced together, the fields will overlap, which is not a problem until the particles are forced too close to each other. Because different particles have different tension values, spin rates, and other spin characteristics, their closely overlapping fields will create an area of turbulence that effectively eliminates or collapses the usability of that portion of the field.

A detection device at each slit may cause the fields of the respective particles to be disturbed to the point of creating turbulence that effectively collapses much of the field. Consequently, the field may not split and diffract, or the two halves of the altered field cannot effectively interfere with each other, and no interference pattern is displayed.

1.2b The Pilot-Wave Theory

Again, particles are always observed as particles, never as waves. This is because they are only particles, but their existence inside the bubble-like, pulsating field makes the signature of their presence wavelike when testing for waves. Physicist Louis de Broglie in the 1920's, and David Bohm in the 1950's, nearly figured that out. They advanced a theory that particles are guided by, or riding on, *pilot-waves*. The pilot-wave theory describes everything quantum mechanics does and explains a good deal more.[10.] The theory assures that particles are both real and present whether observed or not, in opposition to then popular arguments to the contrary.[11.]

Bohm thought the pilot wave was a vibration in some pervasive and sensitive field he called the *quantum potential*. According to Physicist Philip Ball there is nothing obviously impossible about the idea of a quantum potential, it's just that there is no evidence for it.[12.] It sounds like the quantum potential Bohm had in mind is very much like our IM. Initial experiments looked promising but could not be duplicated, so the de Broglie-Bohm theory remains outside the mainstream of quantum physics.

In conclusion, a particle's personal field is simply a disturbance in the fluid-like nature of the IM due to the spin characteristics of the particle. Given that the Double-Slit experiment can be reasonably interpreted as the particle's field passing through both slits and being diffracted and interfering with itself, we can infer that particles are not waves but simply behave with wave-like features.

Prediction: Someday a physicist will extrapolate Bohm's mathematics for pilot waves into the physics of personal particle fields, which will explain the double-slit interference pattern accordingly and resolve the issue of wave-particle duality. That recognition might be shared with the equally enterprising physicist that devises and carries out the experiment that proves that it is the particle's field that is passing through both slits and interfering with itself.

1.3 Experimental Predictions and Rational Conclusions

Once shown that it is the particle's field that is passing through both slits causing the interference pattern, it will mean that particles are real and not waves. It will imply that the existence of a dark energy type intergalactic medium is real, and that fields are personal to particles and not ubiquitous throughout the universe.

Resolution of the wave-particle duality issue is not necessarily proof of the hypothesis for this entire model, and if the experiment proves impossible to perform, or does not result as predicted, it does not mean the model is invalid. The model contains several theories, and each must stand on its own, with its own plausibility and observational evidence. A positive predicted outcome to the experiment would mean, however, that many aspects of the early universe are not as we currently believe and that starting over with a fresh perspective is essential. We will discuss the experiment's impact on current theories as we meet them in the course of building our model.

The accuracy of Quantum Mechanics can be attributed to the particle darting within its bubble-like field, making its exact location subject to probability, which QM handles quite well. The proof that particles are not waves will better define the foundation of quantum theory and our perception of reality, which will be examined in Chapter Nine.

For now, let's return to the beginning of our story, to the very instant the universe began. There we will frame the picture by laying the first puzzle pieces representing the initial conditions and from there build a new, solid foundation to our universe.

Resolution of
Cosmological Problems – Ch.1

☐ Explanation of Work Energy
☒ Foundation of Quantum Mechanics
☐ Magnetic Monopole Mystery
☐ Missing Anti-Matter Mystery
☐ Neutrino Oscillation
☐ Origin of Cosmic Web
☐ Origin of Dark Energy
☐ Origin of Dark Matter
☐ Origin of EM field
☐ Origin of EM Waves
☒ Origin of Fields
☐ Origin of First Stars
☐ Origin of Force
☐ Origin of Gravity
☐ Origin of Mass
☐ Particle Decay
☐ Particle Evolution Sequence
☒ Particle Tunneling
☐ Purpose of Entropy
☐ SMBH Creation
☐ Space & Time Relativity
☐ The Fine-Tuning Problem
☐ The Flatness Problem
☐ The Hierarchy Problem
☐ The Horizon Problem
☐ The Magnetic Field
☐ The Measurement Problem
☐ The Reality Problem
☐ Virtual Particles
☒ Wave-Particle Duality Resolution
☐ What Drives Cosmic Evolution
☐ What is Energy
☐ Why Gravity so Weak
☐ Why Photons are Massless
☐ Why Three Generations

Penergy cannot be looked upon as a fuel that powers things to move; relationships power things to move.

Chapter Two

2.0 Choosing the Initial Conditions

Our goal is to create a new vision of the early universe that is mathematically sound, provides evolutionary continuity, includes known observational evidence, and answers many of the outstanding cosmological mysteries. If we are missing something big in our telling of the cosmic story and we can't see looking back just where we went wrong, then our best approach is simply to start over. Consequently, we are going back to the beginning of the universe. In keeping with Hawking's conclusion discussed in the Introduction, nothing yet has emerged, so the only thing that exists is a spinning speck of pure energy, dubbed Penergy. There is no mass, no space, no fields, no forces, no math, no laws. Until the Penergy changes into something we can identify, there is nothing to measure or find a relationship with. All those things mentioned will *emerge* as the Penergy evolves.

That mysterious, even mystical, Penergy is the substance of everything we know to exist. The history of our universe should be told in terms of *the evolution of Penergy*, but the story isn't told that way. The complex role Penergy played in creating the foundation of our early universe from which everything naturally evolved should be at the heart of the story, but it isn't. The active role Penergy continues to play in cosmic evolution should be a highlight in the story, but it's never mentioned. I fear that our failure to recognize the comprehensive role and dynamics of Penergy in the story of the early universe has resulted in a cohesive, complex vision, but not the real story.

The current story posits several hypotheticals as part of the initial assumptions, such as a flaton field, inflationary expansion, ubiquitous particle fields, and unified forces. For the past hundred years we have been on the lookout for observational evidence to support those hypotheticals, but they remain for the most part mathematical constructs to fit our theories. Quantum Field Theory (QFT) is undoubtedly a grand mathematical achievement, but it's built on some very questionable assumptions. We may have the math right, but perhaps that same math can be utilized under different assumptions that prescribe a more plausible story. Physicist Harry Cliff tells us, "This is a moment to reexamine our assumptions and look at old problems from a different angle. More than anything, it is time to put our grand ideas and preconceptions to one side and listen carefully to what nature is saying."[1] Given our lack of progress in solving major cosmological mysteries and the apparent need to start over, it's clearly time to re-evaluate our initial assumptions.

2.1 Initial Conditions

Many authors see the universe starting from a single point, and all the energy that exists in the universe today existed at that point - a *singularity* of infinite energy and density.[2] Others say that anything infinite does not physically exist. So, why would many scientists still envision a point of infinite energy and density at the start of the universe?

2.1a The GR & Singularity Problem

Within weeks of Einstein publishing his theory of General Relativity in 1915, Physicist Karl Schwarzschild used Einstein's equations to demonstrate that an object in space with sufficient gravity could create something we now call a blackhole, with a core having a singular state of infinite density – what is now called a singularity. In the 1970's, Physicist Stephen Hawking and Mathematician Roger Penrose wrote a theorem concluding that the energy kernel at the start of an expanding universe such as ours, and at the core of all blackholes, is a singularity. Hawking later changed his mind about a singularity at the start of the universe.[3] Yet, the idea of the universe starting from a singularity persists.

Mathematics has many ways to manipulate and deal with infinities, but in physics infinity is a big problem. Physicist Michio Kaku tells us that to a physicist, infinity is just a sign that the equations are not working; that the physicists don't understand what is happening.[4] Einstein believed the presence of singularities was a sign of imperfect physical understanding and that it made no sense for an object to have zero size and infinite mass density.[5] Physicist Paul Sutter informs us that "...singularities don't actually exist; they are flaws in our model of reality."[6]

Although Einstein's equations work perfectly for gravity, they ultimately prescribe a singularity and at the singularity the equations become nonsensical. This presents a dilemma for theorists choosing what conditions to consider at the start of the universe. If we believe that Einstein's equations are accurate and applicable and the singularities exist, then we must admit to the nonsensical aspect of the equations. That suggests we are either wrong about their applicability or there is more to the initial conditions than we are considering.

If we believe singularities don't exist but wish to recognize an applicability of Einstein's equations, we must ignore the singularity or cut the equations off short of the singularity. Here we are choosing to use that part of the equations that suit us and ignore the nonsensical, which seems like a dishonor to these otherwise stellar equations, and suggests we are manipulating the system and risking missing something.

A third option is to sidestep the existence of singularities by not using Einstein's equations initially. Those equations are brought into play when we wind the clock back from the present to the past. Doing so requires keeping track of the conditions of the universe along the way, necessitating Einstein's equations to explain the impact of gravity, but it ultimately brings the singularity into the picture. It is a legitimate use of the equations where gravity is in the picture, but was gravity actually in the picture at the very start of the universe?

It seems that most cosmologists have chosen to use Einstein's equations while ignoring or cutting off the singularity. The reason for this approach is reflected in Dr. Devereux's admission that cosmologists do not like the singularity idea but do not know how else to describe the start of the universe and they put up with it until they find something better.[7.] The problem with that option is, again, if we use Einstein's equations and ignore or cut out the singularity then we run the risk of missing something crucial in our understanding of the early universe. It seems unwise to go down a road that seems mathematically coherent but ends in an abyss that cannot be repaired or circumvented. It brings the quality of the foundation of the entire road under suspicion, undermining our confidence in where it has brought us, or where it is taking us.

Let's address the third option: perhaps Einstein's equations were not meant to be used at the birth of the universe. As noted in the Introduction, Stephen Hawking's final conclusion was that *the laws of physics are not fundamental but are emergent.* Thus, one could argue that employing Einstein's equations at the start of the universe may not be valid. Further, the absence of gravity at the start of the universe seems likely since the existence of gravity in the presence of all that condensed energy should never have allowed that energy to expand.

With nothing yet having evolved at the start of Hawking's universe, not even the emergent laws of physics, it limits the initial conditions we can posit. We are reduced to speculating on the speck of Penergy we believe to have existed. Given the arguments against using Einstein's equations and its singularity, we might posit our initial conditions to simply be that *the universe started from a finite speck of Penergy with a finite density that began expanding*. Given Hawking's final conclusion, these seem to be the only reasonable assumptions we can make. Being restricted to the initial conditions being all about Penergy will change the story. It provides, however, hope of finding an alternative theory that eliminates the need for fine tuning, so many hypotheticals, and positing things into existence, while preserving the basic observational evidence supporting the BBT.

Winding the clock backwards incidental to the development of the BBT seemed like a good idea at the time, but it may have led us astray, so let's not go there. We know the universe is expanding, so we can legitimately presume it was once smaller and denser. We can simply start the universe at time=0 and wind forward with a blank slate and a mere speck of condensed energy that begins expanding. We don't need to resolve how big or dense the speck was but agree it was very dense; a condition I like to call its NIB state - **N**ear **I**nfinite **B**ut (not quite).

If the universe began as a finite amount of energy with a finite degree of density and was free of the equations resulting in a singularity, the foundation of our road leading away from time=0 would be seemingly uncertain but perhaps more secure. While retaining, and perhaps re-evaluating, the observational evidence from the early universe, this new story could give us new theories as to the origin of SMBHs, the cosmic web, dark energy, and more. And perhaps do so while requiring less fine tuning and still provide a seamless sequence to the evolution of fields, forces, and complex matter.

The BBT requiring several exotic hypotheticals in order to work apparently got us off on the wrong foot. Consequently, any replacement theory should be guided by these two fundamental principles: 1) Everything that exists in our universe emanated from an initial quantity of pure energy, and 2) Everything that has and will come into existence in our universe since time=0 has *evolved* from that pure energy with a direct causal connection to a pre-existing condition. That means that we should not allow things to pop into existence, invent new kinds of energy, or turn that pure energy into something else, without a reasonable causal explanation. If we allow those things to happen without a reasonable evolutionary basis, we are again treading a tenuous path and potentially building a house of cards. Demonstrating a clean evolutionary sequence would give the new story a solid foundation.

We are seeking a new beginning to the story that naturally details our cosmic evolution. Starting simply with a speck of expanding Penergy and no other preconceived conditions is restrictive and is a substantial departure from current theory, but it's a clean approach that seems natural and shows promise. It serves as our initial puzzle pieces, whose flexible edges provide a variety of possible connections. The prospect of putting together a new picture that demonstrates evolutionary support, is mathematically sound, and solves outstanding cosmological mysteries is intriguing. First, however, to reframe our picture effectively we must ensure we understand this stuff called *energy*.

2.2 What is Energy?

The predicted outcome of the experiment suggested in Chapter One will reveal that particles are physical things, not waves, making Photons particles, not simply energy waves. This realization requires our looking at energy from a slightly different perspective. The substance of energy itself remains allusive. We don't really know much more about it than when Physicist Richard Feynman told us in the 1960's, "It is important to realize that in physics today, we have no knowledge of what energy *is.*" [8.] That may be true, but we know enough about energy to know how to work with it.

The content and mechanics of our universe are all about energy; the physical energy that defines it and the work energy that drives it. To understand how our universe evolved, we must understand the energy from which it is made. We have identified two classes of energy: *physical energy* and *work energy*. To better understand energy, we must understand both the difference and relationship between the two.

2.2a Physical Energy

We don't usually think of energy as something physical, but we live in a physical world and according to Einstein's $E=MC^2$, energy is equivalent to mass, and mass is the measure of physical matter. Consequently, if we are going to call matter a physical thing, then we must say that the energy constituting mass (*notwithstanding QFT's interpretation*) is physical. That physical energy is usually referred to as mass energy or rest energy, but sometimes it is called pure energy, so to differentiate it from other forms, I will refer to it as *Penergy*. So, to be clear, Penergy is the content of the speck of physical energy from which the universe began and is the substance from which everything in the universe evolved.

2.2b Work Energy

In the science community the definition of energy is *the ability to do work*. Work is the generic description for the transference of energy between various systems regarding the movements and relationships between particles, fields, and forces. Work energy exists in many forms: radiation, light, heat, thermal, chemical, nuclear, kinetic, gravitational, and potential. Let's look at these work energies a little closer to identify common traits.

Thermal and *heat* energy, whether transferred through a medium of a solid, liquid, or gas, comes down to a high-energy electron emitting a photon, which is received by a lower-energy electron (*or vice versa*), thus spreading the level of energy throughout the medium until an equilibrium is reached. Let's label this phenomenon of energy transferred through photons as **Photon Energy**. *Light* energy and *Radiation* energy are also Photon Energy by definition. *Nuclear* energy is also Photon Energy because the energy released by the nucleus in fission, fusion, or decay quickly resolves into particles and photons.

Gravitational energy is derived from a mass curving spacetime, creating an attractive *field* that curves the pathways around it toward the mass. Since it derives from the mass's field, let's label this phenomena **Field Energy**. *Chemical* energy is also Field Energy because it is based on the attractive/repulsive force of the electric field. That force affects the behavior of electrons in atomic/molecular interactions, which we call chemical reactions. *Potential* energy possesses the *potential* for influencing the movement of a body due to its position within a field, either gravitational, magnetic, or electric. Consequently, it too is Field Energy.

Elastic and spring energies are Field Energy because the force holding the molecules that determines the flexibility of the spring is an electromagnetic field. The energy associated with gravitational waves created by such things as bound neutron stars is Field Energy because the movement of the masses creating the waves in the inter-galactic medium is due to their respective gravitational fields. Kinetic Energy is the work energy related to movement.

We've narrowed our work energies down to three categories: Photon Energy, Field Energy, and Kinetic Energy. Let's examine them further.

2.2c Photon Energy

Photons can be absorbed and emitted by charged particles. They are the energy currency used by a particle, atom, or molecule allowing it to attach to, decouple from, or otherwise interact with another particle, atom, or molecule. When a matter particle absorbs a photon, it changes both the velocity of the particle and the calculation for the particle's total energy. The value of the transferred energy is measured in a change in the particle's Kinetic Energy. Photons are a source of work energy because they can be absorbed and emitted, changing a particle's Kinetic Energy.

2.2d Field Energy

A particle's velocity can be changed when acted upon by a force. Forces are derived from gravitational, magnetic, and electric fields (*How forces are derived from fields will be examined in Section 5.2.*) Near these fields an unincumbered particle will be *accelerated*, meaning changed in its velocity, either its speed or direction. The value of the transferred energy is measured in the change in the particle's Kinetic Energy. Fields are a source of work energy because they create a force with the capacity to change a particle's Kinetic Energy.

2.2e Kinetic Energy

A body or a wave when in motion is subject to Newton's First Law: a body in motion will stay in motion until acted upon by an outside force. That means that the unaccelerated motion of a body does not depend on the presence of any intrinsic energy; it needs no additional energy to continue moving. The motion energy a moving body is said to possess is simply the work energy transferred to it by the force(s) that placed it in motion. The body or wave in motion is said to possess energy only due to its capacity to change the acceleration of another body by transferring some or all its energy by collision.

Kinetic Energy of a moving mass is work energy because it has the capacity to change the Kinetic Energy of another mass, but in doing so it gives up its own motion energy. Consequently, Kinetic Energy is not a source of work energy but simply a vehicle for transferring the work energy it possesses that originated from either photon absorption or an encounter with a field.

2.2f Energy Summary

From our Fundamental Principles mentioned earlier we know that every physical thing we know of in the universe evolved from the initial speck of condensed energy (*Penergy*). That is the same energy we know as the Mass or Rest Energy that appears in Einstein's $E=MC^2$, and the energy that manifested itself into what we call particles, dark matter (*if it exists*), and dark energy. Any other energy is a form of work energy. *

Energy is customarily defined as the ability to do work, which is the ability to exert a force that causes things to move, affecting their acceleration. Work energy to affect a force is derived either through

---

* I acknowledge those who believe *consciousness* may be its own energy, but that metaphysical subject is outside the scope of this treatise, so it will have to be addressed elsewhere.

the absorption of a photon or an encounter with a field. We have mathematically found a way to measure Mass Energy and Work Energy in the same terms, but the two energies are different: Mass Energy is related to the content of particles, while Work Energy is related to the behavior of particles.

In conclusion, we observe work energy in the form of photons and fields. Photons, though massless, are nonetheless particles and like all other particles are created from Penergy. (*An explanation for how photon particles can be massless is found in Appendix#1*). Fields are created from the relationship between those spinning particles and the thin-density medium of Penergy. Consequently, the speck of Penergy that started the universe is ultimately the source of all matter energy and forms of work energy.

Our major scientific accomplishments over the past two hundred years are all related to the functioning and working relationships between these three forms of energy – field energy, photon energy, and Penergy. Einstein and Maxwell gave us concise and exquisite equations that describe how gravitational and electromagnetic *field energy* affects matter. Einstein's Photoelectric Effect and the later development of QED describe how *photon energy* interacts with matter. Einstein's Special Relativity defined the relationship between mass and *Penergy* through $E=MC^2$. These cornerstones to understanding our universe are for the most part simply a description of the dynamics of the three sources of energy.

If we were to compose a schematic of our mechanical universe, Penergy would be the substance of its gears, Photons and Fields the energy sources driving those gears, and cosmic evolution the process arranging those gears. Our Standard Models would explain the inventory, and Quantum Mechanics would provide the mathematical formulation giving us our best guess as to the underlying behavior of that inventory at the quantum level. Lastly, the above-mentioned scientific accomplishments would provide instructions as to how it all works. That schematic portrays the skeleton of our cosmic picture puzzle.

2.3 An Overview of Penergy

In our new picture of the early universe, Penergy is neither changed into an *inflaton field* nor consumed in the creation of particles, fields, and forces within the first second of a big bang. It is simply a speck of Penergy that is allowed to expand and evolve. Think of it as a near-frozen fluid, gradually thawing and expanding through various phases from a heavy goo to a very light density fog. Over the past 13.8 billion years, the speck has lost nearly all its density, expanded to enormous proportions, and smoothed out into a thin-density, fluid-like, intergalactic (*dark energy type*) medium, leaving supermassive blackholes and particles in its wake.

Penergy is the only real substance in our entire universe and it appears in many manifestations. It is difficult to define or imagine because there is nothing else like it. It is referred to herein as a cloud and a bubble, but that is only to give the reader a visual image. It's not solid, liquid, or gas, but it has the capacity to make itself into those things. It is very dynamic, meaning it has the capacity to be many things (*blackholes and particles*); to make things *happen* (*evolution and forces*); to bring about non-tangible things (*consciousness and intelligence*); and to possess even more capabilities we are only beginning to realize.

Penergy has a few observable traits. It is a very dynamic, elastic substance that *moves, expands, spins*, and *contracts* depending on its density. Any energy attributed to a process involving one or more of these traits should be considered as derived from Penergy's natural attributes.

We know energy moves; it is practically the definition of energy. Nothing is perfectly still; everything jitters because energy moves.

I imagine the Penergy speck to have once been a stable, fast-spinning, very compact speck that for reasons unknown lost its stability and began expanding. Since we know that space defining our universe has been expanding since its birth, we can infer that Penergy, when not in its compact, speck state, naturally expands.

Everything in the universe spins: galaxies spin, blackholes spin, stars spin, planets spin, particles spin, and in fact everything in the universe spins or is made from something spinning. This strongly suggests that spin is also a natural trait of Penergy. I imagine the initial speck to be spinning at a very high rate when it began expanding. That angular momentum would have been diffused throughout the Penergy medium, gradually reducing in intensity as the universe expanded. Consequently, the entire early Penergy medium would have been spinning and would have retained spin characteristics causing newly formed objects to spin. It would have encouraged large swirling clumps of Penergy to break away and form independent entities. Consequently, the angular momentum of SMBHs is not only generated from within but is helped by the churning of the early Penergy medium itself and may even possibly influence the rotation speeds of galaxies and clusters of galaxies created in the early universe.

Penergy also seems to have the natural tendency to contract. Spin will cause fluids and solids to expand outward, but in the case of Penergy, which remember is pure energy and not like any other substance we are acquainted with, I believe its spin creates a tension in the elastic, Penergy medium that causes a contraction.

A large body of cosmic gas with optimum density will begin to swirl on its own accord and as it swirls it will draw in more of the surrounding gas, a process called accretion. Experiments reveal that rotating objects display an inward acceleration that is customarily ascribed to gravity, but in the early process of accretion, gases are seemingly drawn in before a commensurate gravitational field can be established. What appears like gravity may actually be a *contraction* in the surrounding medium initiated solely by spin. Physicist Michio Kaku explains that according to General Relativity, the faster you move the flatter you are because space compresses. For something spinning, the outer edge is spinning faster than the interior, so the rim contracts more than the interior since it is moving faster. [9.] This may be observational evidence for spin causing contraction.

Stars initiate the spinning process of accretion on their own accord, so there is little reason to doubt that Penergy can do the same when in an appropriate density state. Consequently, it is reasonable to posit that when Penergy is in a light to medium density state it will begin to swirl, spin, and contract, pulling in the surrounding Penergy, creating a denser and denser body in the same way the mass of a star assimilates.

Penergy cannot be looked upon as a fuel that powers things to move. Penergy is merely a substance with definable traits that allows for the creation of things and relationships with and between those things. Those relationships allow for movement, attraction, and repulsion. Penergy doesn't power things to move; relationships power things to move.

Penergy behavior is a function of its density, which will be the subject of much of our study. We can address the relative densities of Penergy by creating a simple density scale. If the near infinite density, NIB-state, was a 1.0, we can break down the lesser densities by .1, making them .9, .8, .7, etc. The scale will help us understand Penergy density in relative terms, but the exact densities we talk about are only guesses.

We have reframed our cosmic picture – it's all about energy. Mathematically, we are starting from scratch, so the first puzzle pieces do not have to conform to a certain fit. We can allow the Penergy to change and evolve and see what shape it takes. We can wait and see what the puzzle is telling us! We will ultimately recognize those initial puzzle pieces to be the first things that formed out of the heavy but thinning Penergy – they are largest individual things to exist in our universe.

Resolution of
Cosmological Issues - Ch. 2

☒ Explanation of Work Energy
☒ Foundation of Quantum Mechanics
☐ Magnetic Monopole Mystery
☐ Missing Anti-Matter Mystery
☐ Neutrino Oscillation
☐ Origin of Cosmic Web
☐ Origin of Dark Energy
☐ Origin of Dark Matter
☐ Origin of EM field
☐ Origin of EM Waves
☒ Origin of Fields
☐ Origin of First Stars 3
☐ Origin of Force
☐ Origin of Gravity
☐ Origin of Mass
☐ Particle Decay
☐ Particle Evolution Sequence
☒ Particle Tunneling
☐ Purpose of Entropy
☐ SMBH Creation
☐ Space & Time Relativity
☐ The Fine-Tuning Problem
☐ The Flatness Problem
☐ The Hierarchy Problem
☐ The Horizon Problem
☐ The Magnetic Field
☐ The Measurement Problem
☐ The Reality Problem
☐ Virtual Particles
☒ Wave-Particle Duality Resolution
☐ What Drives Cosmic Evolution
☒ What is Energy
☐ Why Gravity so Weak
☐ Why Photons are Massless
☐ Why Three Generations

*The disappearance of the intermediate size
swirls at the end of the cascade is the reason
we see very large voids in the cosmic structure,
and it is the reason we see very few SMBHs
sized below fifty thousand solar masses.*

Chapter Three
3.0 The Origin of Supermassive Blackholes

The hypothesis again is that the universe started from a speck of pure energy, deemed Penergy, that expanded through various phases and transitions, settling into a smooth, light-density medium that defines our universe. We will next examine those transitions, exploring what may have naturally evolved from that spinning speck. Our exploration, supported by observational evidence, will reveal that the first things to have evolved, making them the oldest things in the universe, were supermassive blackholes (SMBHs).

A blackhole can be created when a star several times larger than our sun reaches the end of its life and is theorized to implode into a much smaller, denser mass that goes through various stages, condensing further and further, eventually becoming a blackhole. The stellar collapse into a blackhole is believed to be the normal life cycle of stars that are twenty to one hundred times bigger than our sun. SMBHs, however, are millions of times larger than our sun and are created much differently.

3.1 Blackholes

Many authors refer to the heart of a blackhole as a singularity. As discussed in Section 2.1a, singularities arise from Einstein's equations on gravity, but the equations become nonsensical at the singularity. That makes those mathematically constructed but otherwise hypothetical singularities at the center of all blackholes problematic. Singularities tell us that either the equations are not applicable, or we are missing something. There is no reasonable compromise that would allow for singularities, so let's not go there. So, what is at the center of a blackhole if not a singularity?

In describing how nothing escapes a blackhole, Stephen Hawking wrote, "Thus the matter inside a blackhole would be trapped and would collapse to some unknown state of very high density."[1.] Matter crushed to a very high density would likely turn back into Penergy, which would then constitute the bulk of the blackhole's content. Accordingly, we might theorize that any particle mass drawn inside a blackhole is being crushed or otherwise processed back into an amorphous state of Penergy, and the Penergy itself is being condensed further and further into a very high density state, perhaps back to its NIB-state, as defined in 2.1a..

Consequently, a blackhole does not need a mass of particles to come into existence or sustain itself. The Schwarzschild solution to Einstein's field equations dealing with the creation of blackholes does not require the presence of matter.[2.] All of this suggests that the mass of a blackhole is more about condensed Penergy than about particle matter, a fact that will help us understand how SMBHs were created.

At the heart of all galaxies is a SMBH, as many as billions of times bigger than our sun. Scientists have not yet agreed on a sound theory for SMBH creation. They are hamstrung trying to develop a theory for SMBH creation after atoms were created in keeping with the Standard Model's version of the Big Bang Theory (BBT). Trying to base a SMBH creation theory on the Hot Big Bang platform doesn't work well absent layers of hypothetical conditions. Blackhole growth simply by accretion is slow, taking as much as two billion years to grow to supermassive size.[3.] Computer simulations have indicated possible complex processes involving a mixture of collapsing gas formations, temperature differentials, density fluctuations and perturbations, all of which must be just right to form large galaxies. Such a complex scenario might be capable of producing large galaxies, but not likely the billions of galaxies with an attendant SMBH at their center.

Under the Hot Big Bang Theory, the most feasible way for SMBHs to have grown to billions of solar masses is through blackhole and galaxy mergers. Let's examine that possibility.

3.2 Arguments For and Against SMBH Creation by Mergers

The Hubble and James Webb Space Telescopes (JWSTs) have produced deep-space images showing galaxies that have been calculated to have formed well within the first billion years after the big bang. Such a short period does not allow enough time for stars to form, die, blackholes to form, and for the blackholes to have grown to SMBH size simply from isolated mergers.[4.] The merger theory might work, however, if the merging blackholes were uncharacteristically large.

Astronomer Martin Rees developed a theory that larger blackholes could have started from seed blackholes created by the collapse of thousands of stars, and then those larger blackholes could have merged to create SMBHs.[5.] Scientists are now speculating on seed masses of 100,000 solar masses produced from the collapse of early giant gas clouds. Those complex scenarios may have happened somewhere in the universe, but again it doesn't seem plausible for the creation of every SMBH at the heart of billions of galaxies.

Computer simulations have been tweaked to the point of allowing very large blackholes to have formed in the early universe. But all those simulations eventually require mergers to take place for

SMBHs to be created within a billion years after the big bang. Mergers at that age of the universe would require whole galaxies to have merged over and over to grow to a SMBH size. But observational evidence regarding spiral galaxies with SMBHs suggests their formation could not have happened through a frenzy of galaxy mergers. A spiral galaxy has a center mass with arms growing out around it, as shown in Figure 3.1.

Figure 3.1 Spiral Galaxy

It is believed that when spiral galaxies collide and their black-holes merge, the process tends to scatter the spiral-disk shape.[6.] With multiple mergers, the consolidation would create a diffused elliptical galaxy, as shown in Figure 3.2.

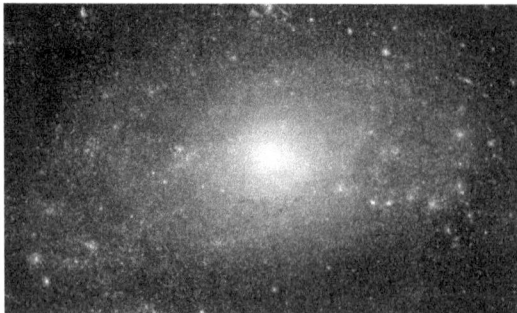

Figure 3.2 Elliptical Galaxy

Seventy percent of the galaxies in our universe are spiral in shape.[7] That we can observe so many pristine spiral galaxies containing a supermassive blackhole suggests those galaxies have not suffered the turmoil of multiple mergers and apparently acquired their supermassive blackhole by other means. The remaining thirty percent of galaxies are elliptical and generally house a much bigger supermassive blackhole billions of times larger than our sun. They look like a near spherical cloud, and due to their size and appearance they indeed could be the remnants of galaxy mergers. The Milky Way is expected to become an elliptical galaxy once it collides with the Andromeda galaxy billions of years from now.[8]

If SMBHs were not formed by either accretion, collapsing gas clouds, or mergers, how were they created? Recall that Penergy naturally moves, expands, spins, and contracts. Given that the initial speck of Penergy constantly thinned in density as the universe expanded, there is a very plausible explanation for how those supermassive blackholes came about. The short answer is that they could have formed out of Penergy when it was quite dense in the early universe, incidental to a process called the *cascade effect*. Let's examine how that may have happened.

3.3 Alternative Theory for SMBH Creation

We can only guess how an expanding speck of pure energy would likely evolve. Pure energy is the physical substance from which we are made, yet for us it is largely intangible and remains a mystery. Our observational evidence is that it moves, expands, spins, and contracts, and using those traits we can only put together a plausible theory for how it might have evolved, but it's only a guess. We will strive for a theory that is within the bounds of our fundamental principles, is plausible, and fits with the observable universe we see today. At this early stage, the substance we are calling Penergy, and the conditions in which it exists, are so foreign to what we have dealt with so far that we don't even have much in the way of solid mathematics to work with.

Such a theory must account for the development of large galaxies forming so early, the tremendous size of the SMBHs at the heart of those galaxies, and the large-scale structure of the galaxies we call the cosmic web. The following alternative for SMBH creation accounts for all of that, but other theories may be equally plausible.

It is difficult to say exactly what transitions the expanding Penergy would have gone through on its journey to become what it is today. As the spinning Penergy expanded and lost density, its initial state would likely be swirling and dense. As it expanded and thinned, the swirls would have dampened out, but in its very early stages, I imagine the Penergy to be a roiling, thick dark fog.

As observed earlier, blackholes, stars, and planets spin, and if stars can start the process of their own creation by swirling the substance from which they are made, Penergy should be capable of doing the same. Consequently, as the expanding Penergy lost density it would naturally begin to move, churn, and at some optimum density, say .7, it could have separated into huge independent swirls, drawing from the angular momentum inherent in the initial roiling medium. This is very much like the way we understand stars to begin. Astronomer John Barrow tells us, "Studies of compressed [gas] clouds suggest they will become unstable, collapse, and eventually break apart. Finally, stars form out of the self-gravitating gas fragments."[9.] My hypothesis is that the creation of SMBHs occurs much the same way.

The developing swirls would be large and quite heavy so the movement would initially be slow. Eons of time and expansion could have taken place before there was an otherwise noticeable change. But eventually, like star formation, as the circular motion gradually sped up it would have gained a growing gravitational-type compression causing the Penergy to pull in toward the center of the swirls, leaving thinner-density areas at their outer edges.

At those outer edges, now at say a .6 density, many smaller globs would swirl into existence and begin drawing in available Penergy, leaving their outer edges less dense. Those outer edges, now at say .5 density, would have created many smaller swirls, again with their outer edges thinning, setting off a cascade of swirls of lesser and lesser density, a process coined the Cascade Effect, as illustrated in Figure 3.3. Of course, the interval between swirl creations could have been .05 or .01 density, it is only a guess.

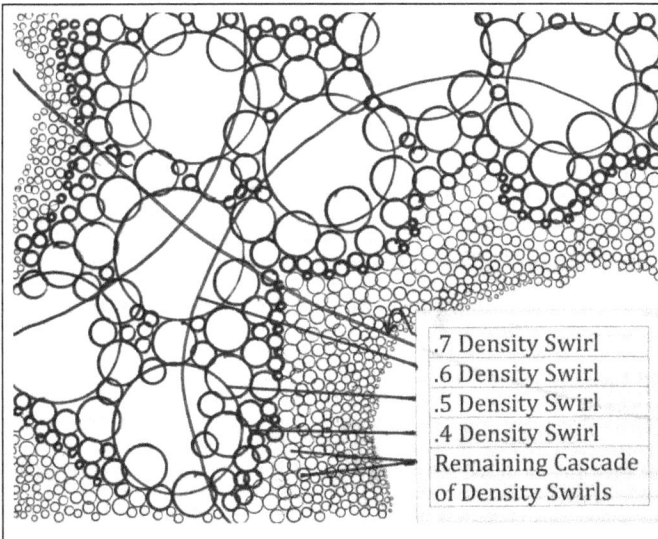

.7 Density Swirl
.6 Density Swirl
.5 Density Swirl
.4 Density Swirl
Remaining Cascade
of Density Swirls

Figure 3.3 The Cascade Effect, showing how each swirl (*depicted by circles*) drew in Penergy, causing its outer edge to thin in density, allowing new, smaller swirls to begin. This effect happened over and over, cascading through smaller and smaller swirls with diminishing densities.

The expanding thin density Penergy would eventually produce billions of swirls of different sizes from huge to tiny, causing all kinds of anomalies in their distribution. As the swirls became better defined, they would have begun to create a stronger gravitational influence that would have slowed the Penergy expansion near them in proportion to their density, causing even more anomalies.

In time, the inexorable expansion would have separated the swirls into distinct entities. The larger swirls would continue swirling and contracting, gaining angular momentum with a growing gravitational influence. Eventually the swirls with sufficient density and angular momentum to sustain themselves would become *pure energy blackholes*. Those blackholes, many million times the size of our sun, would eventually be at the core of billions of galaxies making up our universe. The fluctuations, contractions, territorial battles for Penergy, and the *cascade effect*, would cause the resulting galaxies to form in clusters, groups, walls, filaments, and other peculiar distributions that would ultimately shape our universe.

Pure energy blackholes are not yet recognized but this scenario, or one like it, seems to be a credible resolution to the mystery of their formation. An article in *Quanta Magazine* in December 2017, entitled *Earliest Blackhole Gives Rare Glimpse of Ancient Universe,* discussed the discovery of a very old blackhole with its age placed at 690 million years after the big bang and its size equal to 780 million times the mass of our sun. The researchers' calculations discovered that if the blackhole started out as a collapsed star like a blackhole is normally formed, it would not have had time to grow to its enormous size. They further calculated that if it came into existence soon after the big bang, it would have had to start out about 1000 times the mass of our sun.

The researchers speculated that perhaps hydrogen/helium gases in the early universe condensed into a blackhole, though admittedly this is inconsistent with scientists' current view of the early universe. Since then, scientists have found galactic blackholes measuring several billion times the size of our sun. The JWST is also finding a greater number of large galaxies than expected, which is quite consistent with the cascade theory. Given these reports, the creation of early, pure-energy blackholes seems very plausible.

Under this scenario, we would have a universe largely comprised of emerging SMBHs originating from medium density Penergy. At the end of the cascade of swirls the Penergy density would have thinned to a point where *early elementary particles* could spin into existence. As the universe expanded and thinned further it would have left a very thin density residue of Penergy that now serves as our intergalactic medium. We will soon examine the arguments for the creation of early elementary particles and the intergalactic medium, but first let's look at the process that eventually formed the cosmic web.

The Cascade Effect would have created many swirls of various sizes that spun into existence, but we only see very large supermassive blackholes and very small, tightly wound particles. What happened to all the swirls sized in between the two?

3.4 Spin/Density Relationship with the Penergy Medium

We have observed that when particles heavier than our subatomic particles (*Sub-A's*) are created in colliders and accelerators, they are unstable and immediately decay into smaller, stable particles. This happens consistently and without exception. It implies that those heavier particles are unstable in the thin Penergy density level currently pervading our universe and are forced to decay. A similar fate likely met the medium-to-small sized swirls in the early universe.

As the universe expanded and the density of the medium grew thinner, it would draw off the outer layers of the swirls causing a gradual disintegration. A swirl could only become stable when its gravity was strong enough to hold the swirling cloud together, overcoming that disintegration. The strength of the swirl's gravity was of course a measure of its mass, which was in proportion to the density at which it was created and the density of its core. The larger swirls would have had time to begin condensing their inner-core Penergy into a higher density, thus strengthening their gravitational influence. The stronger gravity and angular momentum would give the swirl a spherical definition, stability, and protection for its outer layer, eventually turning the once spinning cloud into a stable SMBH with a stable equilibrium.

With the universe expanding and the intergalactic medium falling in density, the smaller swirls, say at .4 density or less, would have had neither the initial density nor the time to condense their inner core to a high enough density to give them enough gravity for stability. As the surrounding Penergy density dropped, the medium-to-small sized swirls could not generate enough gravity before their outer layers disintegrated, so the swirls between say .4 and .15 density gradually disappeared, leaving huge voids.

It is fortunate that the density of swirls between .4 and .15 (*or some such range*) were unstable. Had all the swirls developed into blackholes, the total gravitational energy in the universe would likely have brought the expansion to a halt and may even have reversed it. Fortunately, the deteriorating swirls put enough thin density Penergy back into the medium to keep it expanding. As we see in many areas, the universe seems to have a knack for balancing things out, maintaining reasonable equilibriums throughout.

The falling density of the Penergy medium apparently slowed and levelled off at some point, let's say at .04 density (*but it could be .004, it's only a guess used for reference*). That made disintegration less of an issue for the tiny, particle sized swirls because at this stage there was not a wide difference between particle content and medium content. Particles created at some small density, say .15 would not simply be swirling but would be spinning. They would have found the right combination of spin-rate to mass ratio to give them stability, as calculated in Section 1.3a. At this point, to conserve angular momentum, particle production would be in the form of particle pairs wherein two particles are spun into existence simultaneously, one spinning left and one spinning right.

Those stable, .15 early elementary particles would have had the capacity to combine but at the same time would have been subject to annihilation, requiring them to go through an evolutionary phase to determine the best form for combining and possessing a strong coupling strength. They would have begun combining into first-generation, heavy, composite particles, like we see in the Top Quarks and Tau Leptons. The forces holding those heavy, complex first composites together were in an equilibrium state created between the spin of the particles, the denser Penergy medium, and the fields created by those spinning particles. But that equilibrium of state holding the heavy composites together was changed when the density grew thinner. That change in equilibrium caused an instability resulting in a series of phase transitions by decay.

The first heavy composites would have decayed into the second generation, at say 1.0 density, creating Strange Quark and Muon sized particles. They too were apparently too massive and unstable in the thinning density, ultimately decaying at say .05 density into the third generation of quarks and leptons we know today. Again, the stated values of the densities at which all those phase transitions took place are of course only guesses and the actual densities could be quite different. The Standard Model lists those three generations in reverse order.

There were at least two generations of those heavy, unstable, elementary composites, but possibly there were more. The first two generations can, however, be brought back into momentary existence when a knot of Penergy rises sufficiently in density, such as when two protons are smashed together at the Large Hadron Collider (LHC), or when high energy products are created and shot out into space from blackholes.

Those transitions would eventually leave the universe with only three stable elements: various sizes of SMBHs, tightly wound Early Elementary Particles, and a thin density Intergalactic Penergy Medium (IPM). These three constitute all the Penergy that existed in the initial speck from which the universe began. SMBHs, elementary particles, and the IPM are the foundation for our universe, and it is relationships between those three entities from which everything else in our universe evolved, including fields, forces, and complex matter.

3.4a SMBH & Particle Stability Calculation

The universe's creations are structured in symmetries and equilibriums. From all appearances, the universe abhors imbalances and inequalities. In the chaotic environment of the Penergy medium, equilibriums appear essential for structural integrity. SMBH stability is believed to be due to an equilibrium between its traits but the formula for that equilibrium is the subject of debate.

It would seem reasonable for an early elementary particle to also be stabilized by an equilibrium. The particle would likely have some ideal ratio between the energy of its spin and the energy of its mass. I don't know what that ideal ratio would be, but for the sake of example let's make a test case to see how that might work out.

The hypothesis then is that the point at which a massless, early elementary particle spinning on its own axis is stable is when its energy related to its spin was equal to the energy related to its mass. (*How particles can be massless is examined in Appendix #1.*) The energy related to its spin is measured in Planck's equation, $E=hf$. In this equation 'h' is Planck's constant measured at 6.62×10^{-34} Joule Seconds. Joule Seconds is a measure of angular momentum ($kg\, m^2/s$). Plank's constant appears in many quantum theory equations but it is not traditionally thought of as angular momentum. Kg m^2/s is nonetheless the mathematical expression for angular momentum, so I do not believe I am out of line expressing it that way.

The 'f' in the equation is a frequency, measured in hertz, $1/s$. Multiplying the two together, hf gives a value for energy ($kg\, m^2/s^2$), which I call *spin energy.* One might argue that using Planck's constant in this instance is not valid because the frequency the equation is referring to is the particle's wave frequency, not its spin frequency. Recall from Chapter One that the particle is only a particle and not a wave, and that the only way it has a wave frequency is through its spin. I'm simply saying that for a massless particle spinning on its own axis, its wave frequency is the same as its spin frequency. Mass energy is of course measured in Einstein's $E=MC^2$. To test whether at some optimum point these two measures of energy could be the same, we set them equal to each other: $hf=MC^2$.

Let's make up plausible test values: a reasonably high spin rate, say 10^{19} per second, and a possible mass value, say 7.362×10^{-32} kg. The equation $hf=MC^2$ sets up as: 6.62×10^{-34} kg $m^2/s \times 10^{19}$ $s^{-1} = 7.36 \times 10^{-32}$ kg $\times 9 \times 10^{16}$ m^2/s^2. Doing the math, we arrive at equal values on both sides: 6.62×10^{-15} kg $m^2/s^2 = 6.62 \times 10^{-15}$ kg m^2/s^2. Assuming the formulas used for both spin energy and mass energy are valid for a massless, early elementary particle, it appears that it is possible to have a spin rate and mass that allow for the two energy values to be equal, seemingly creating an equilibrium that makes the particle stable.

The actual values for spin rate and mass could be quite different, but the concept would be the same – particles become stable spinning out of a light density medium when some essential ratio is arrived at, and that ratio might be when the energy value of the spin rate is equal to the energy value of the mass. If the particle under creation reached a maximum spin rate before an optimum ratio is reached, the particle might shed mass until the point of the optimum ratio is obtained. Reaching a maximum spin rate and a precise stability ratio made all early elementary particles exactly alike and gave them a definite and precise density and definition, separating them from the surrounding thinner Penergy medium.

Before examining how SMBHs, particles, and the IPM evolved into the universe we see today, let's examine the observational evidence supporting the creation of SMBHs and the Cascade Effect.

3.5 Evidence for SMBH Creation & the Cascade Effect

The hypothesis is that medium density Penergy swirled into pure-energy supermassive blackholes, their distribution was structured by the *cascade effect*, and their development was well underway when the elementary particles and stars evolved. Let's examine the observational evidence supporting that hypothesis.

The Number and Size of SMBHs: SMBHs are known to exist in many sizes, having been measured in hundreds of thousands, millions, and billions of solar masses. The wide range of sizes, all far too large to have been created from solar collapse, is consistent with the cascade effect. The disappearance of the intermediate size swirls at the end of the cascade is the reason we see very large voids in the cosmic structure, and it is the reason we see very few SMBHs sized below fifty thousand solar masses.

The Cascade Effect should have left some very large galaxies created very early. That is exactly what the JWST is finding as reported in numerous articles. Scientists are straining to explain the huge size of galaxies that have come into existence less than 400 million years after the big bang but viewing the early universe from the perspective of a thinning Penergy density and the Cascade Effect, they make perfect sense.

The JWST is making new discoveries weekly regarding the age, size, and distribution of the earliest galaxies. In a November 2023 article published in *Live Science*, Brandon Specker reported that the JWST recently discovered a chain of massive galaxies measuring in the billions of solar masses having been formed about 12 billion years ago. The group has been dubbed "The Vine", and is depicted as a large, bow-shaped array that may be part of a larger cluster [*perhaps a circle*]. The report suggests a circular formation, much like that pictured in Fig. 3.3.

Galactic Distribution: Galaxies are not randomly scattered across the universe but are preferentially found in super clusters, smaller clusters, different sized groups, walls, sheets, and filaments, with very few isolated galaxies. According to Astronomers Luke Barnes and Geraint Lewis, many *dwarf galaxies* have been found in a plane rotating around our Milky Way galaxy, as well as the Andromeda galaxy. A group of similar-sized galaxies rotating around a larger galaxy is the exact description of the Cascade Effect. That phenomenon was thought to be a statistical fluke until additional studies found a similar plane of dwarf galaxies around Andromeda. As Barnes expresses it, "One plane of galaxies might be a coincidence, but two starts to look like a conspiracy,"[10.] A conspiracy in this sense might be an unlikely configuration, but one that makes perfect sense given the Cascade Effect.

The cascading swirls left galaxies embedded in clusters and groups down to the level of dwarf galaxies encircling larger individual galaxies. The Cascade Effect should have left the smaller swirls more prevalent, which bears out. It is reported that dwarf galaxies are the most common galaxy type in the universe. Astronomer Chris Impey describes the galactic terrain as "littered with dwarf galaxies".[11.]

The Cascade Effect should have also left some of the smallest galaxies appearing within the area of the voids. This is exactly what is observed. Astrophysicist Paul Sutter reports that if you zoom in on a portion of the cosmic web and look at the voids you will see dim dwarf galaxies. He goes on to say that upon a close examination of those voids, one sees a faint echo of the cosmic web; "...the cosmic web can be seen as a series of *nested* cosmic webs, each 'level' dimmer and smaller than the last."[12.] That description sounds very much like a cascade of the Cascade Effect, with galaxies forming on the edge of larger galaxies, that are forming on the edge of larger galaxies.

Other huge rings of galaxies have been recently found, as reported in an article posted by Big Think, dated January 23, 2024, entitled *Giant Ring? Giant Arc?* The article reports that the rings of galaxies are so large they do not fit any current model, so many scientists are looking for ways not to believe they exist as independent structures. The article describes a newly found ring of galaxies 4 billion light years in circumference. It is added to the list of seven other galaxy structures that are measured at two to ten billion light years across. This description of arc shaped galaxy formations also sounds very much like what would evolve from the larger swirls in the cascade effect. I expect the JWST to find more evidence of galaxies created in circular clusters, fortifying the argument for the Cascade Effect.

Galactic Movement: Astronomers have observed abnormal rotational speeds for many galaxy clusters, suggesting either additional matter must exist somewhere within the cluster or Einstein's General Relativity needs tweaking. The prevailing theory is that the presence of a hypothetical *dark matter* is adding unseen mass to the cluster, but gravity tweaking theories are out there, as well. The rotational motion of clusters of SMBHs due to the cascade effect, however, may account for some of the observed anomalies.

As Paul Sutter phrases it, "Galaxies are buzzing around in orbits with the cluster."[13.] This observation is consistent with the Cascade Effect. Many of the galaxies were created while their Penergy was in a swirling motion being dragged around by a larger swirl. Given the conservation of their rotational motion, there is no reason to believe those SMBHs that eventually formed would have ceased that circular motion, which could easily be interpreted today as *galaxies buzzing around in orbits.*

Galactic Halos: Astrophysicist John Gribbin reports astronomical observations of *deem blue galaxies,* which are described as very numerous and very faint dwarf galaxies. They existed billions of years ago but have since disappeared as evidenced by there being no younger galaxies of that type. Those faint dwarf galaxies are so numerous they have been described as 'cosmic wallpaper'.[14.] Dr. Gribbin's description describes the swirls very close to the threshold for SMBH stability. They may have existed long enough to form galaxies with stars but ultimately as the Penergy density thinned their SMBH could not sustain itself and would have suffered a forced decay back into the Penergy medium, leaving faint evidence of the galaxy's existence. All of that is consistent with the Cascade Effect.

Galactic Shape: It is observed that both galaxies and solar systems are generally found in the shape of a disk. This may be because they both form in a very similar manner.

Stars form from the accretion process of a high concentration of gas and dust collecting, rotating, and drawing in material until the ball of gas flattens into a disk. Once the star's nuclear fusion begins, the band of residual matter at the outer edge of the star's disk coalesces into planets circling in the same plane creating a solar system.

Similarly, by the time the cascade drew down to the smallest swirls, the large swirls had picked up sufficient angular momentum to cause the adjacent Penergy and newly minted particles and composites to flatten into a disk. As newly created particles combined into atoms and gas, the atoms and gas swirled into stars. Once the SMBHs gained sufficient angular momentum they would have pulled in the surrounding gas, matter, and stars, encircling them in the same plane as the original disk, creating the billions of spiral galaxies we see today.

In summary, just as planets were formed in a plane incidental to star formation, stars were formed in a plane incidental to SMBH formation. This implies that the SMBHs had started forming much earlier but the final phase of creation for both SMBHs and stars was close in time.

Relationship between Galactic Stars and SMBHs: The formation of galaxies is usually attributed to a build-up of matter starting with the formation of atoms, then gases, then stars, then a galaxy of stars. There is little mention of SMBHs as being a part of the galaxy creation process aside from stars dying or huge gas clouds collapsing and forming blackholes that merge until a SMBH is created. It is difficult to imagine how a diffuse creation of stars can die creating a diffuse number of blackholes that then merge over and over in such a way they ultimately create the exquisitely structured spiral galaxies we see today.

Astronomer Chris Impey tells us that every galaxy has a SMBH at its center and the mass of the blackhole is tightly coupled to the mass of the stars in that galaxy.[15.] This observation suggests that perhaps the stars and SMBH were created simultaneously, and once established, the SMBH drew in just the correct number of stars that it could support with its gravitational field. That description seems more plausible for the creation of stable orbits of stars and the creation of spiral type galaxies.

Matching our Cosmic Structure – the Cosmic Web: As observed, galaxies are not spread uniformly across the universe but 2are gathered in clusters, groups, walls, and filaments. The overall arrangement has a honeycomb appearance and is commonly referred to as the Cosmic Web. It is described by Astronomer Wallace H. Tucker, as "massive galaxy clusters as nodes that are interconnected through an intricate web of filaments and sheets of tenuous gas and galaxies, with nearly empty regions called voids taking up most of the volume".[16.] That description is depicted in Figure 3.4.

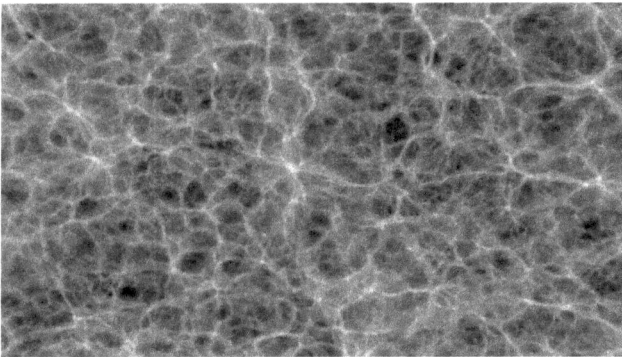

Figure 3.4 Depiction of the cosmic web. Credit: Virgo – Millennium Simulation Project/Springel et al. (2005)/Max Planck Institute for Astrophysics.

The cosmic web formation has been computer simulated resulting in a complex process involving dark matter, oscillations, the birth and death of stars, and galaxy mergers, but why it may have formed that way is still a missing piece.[17.] Let's examine an alternative theory that accounts for the creation of the cosmic web in a more natural and less complicated way.

As illustrated in Figure 3.3, a large .7 density swirl would create smaller .6 density swirls at its outer edge, that would in turn create .5 density swirls at their edges, cascading down to smaller swirls and eventually particles. That process is called the *Cascade Effect*. The larger swirls would eventually become supermassive blackholes. The intermediate and smaller sized swirls, say at .4 density and below, being too weak to sustain themselves would eventually disappear, as detailed earlier. Their disappearance would have left the empty spaces that expanded into the very large voids we see today.

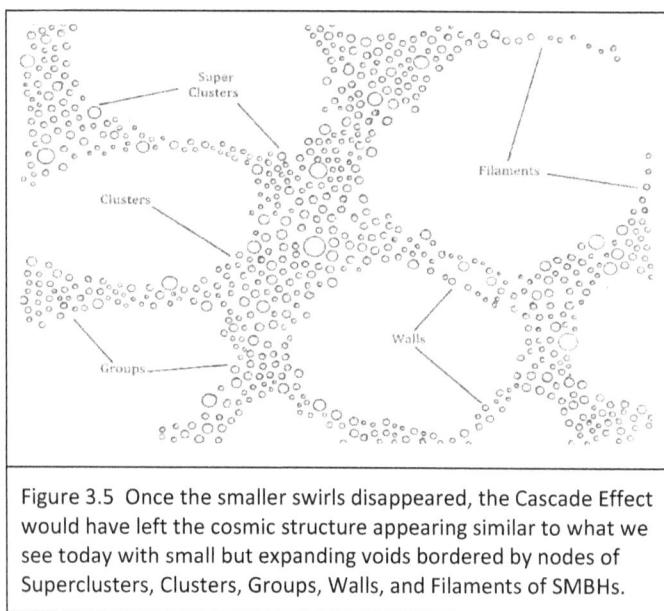

Figure 3.5 Once the smaller swirls disappeared, the Cascade Effect would have left the cosmic structure appearing similar to what we see today with small but expanding voids bordered by nodes of Superclusters, Clusters, Groups, Walls, and Filaments of SMBHs.

The intersection of the larger swirls would settle into super clusters and clusters of SMBHs. The large voids would be bordered by smaller clusters, groups, walls, and filaments, as illustrated in Figure 3.5. Within the clusters, the gravitational influence of the SMBHs would slow the expansion between them, while the voids would expand more rapidly creating even larger voids, giving the overall structure a web-like appearance. Again, the disappearance of the smaller swirls creating the voids is the reason we don't see SMBHs sized less than fifty thousand solar masses.

This observational evidence makes a good argument for SMBHs developing out of the dense Penergy and prior to star formation. When scientists extrapolate their theories on how the universe will ultimately evolve, they often place blackholes as the last things in the universe to exist. This seems appropriate, as it appears they were also the very first things to exist.

Working Prediction: Someday a physicist or astrophysicist will develop the math to properly describe Penergy density. Once that is accomplished, the ability to create a computer simulation of the early universe will likely confirm that SMBHs can be, and probably were, created by some calculated density of Penergy.

Simulations will reveal the high plausibility of pure energy SMBHs, which would confirm that Penergy existed in different densities as the early universe expanded. Simulations will also confirm Penergy's natural tendency to thin and spin, which will explain the Cascade Effect and how different generations of particles were formed.

The observational evidence for SMBH creation and the Cascade Effect seems compelling. I am confident the JWST and future telescopes will continue to find galaxy nesting and early galaxies moving in circular formations, supporting the Cascade Effect. In the absence of simulation confirmation or a better theory, the best rational conclusion will be that SMBHs were created from medium density Penergy, and that SMBHs, particles, and stars came into existence at roughly the same time.

3.5a Origin of the First Stars

As the gravitational influence of SMBHs grew, it would have slowed the rate of cosmic expansion and rate of Penergy density reduction. This would have allowed the smaller swirls that spun into existence to live a bit longer before disintegrating back into the Penergy medium. Many of the smaller swirls could have existed long enough and had sufficient gravitational influence to draw in an appreciable amount of early particle creations. Those swirls would have grown in mass and angular momentum, allowing them to transition from proto-blackholes to proto-stars. Those stars would have been the first to be drawn into the growing gravitational field of a SMBH, creating the first galaxies.

As already detailed, the cascade of swirls would have grown smaller and smaller, eventually spinning into tiny particles. The Penergy medium would have continued to lose density, thinning past the particle creation phase and settling into a thin density Intergalactic Penergy Medium. In Chapter Four we will detail the creation of early elementary particles and address the creation of the Intergalactic Penergy Medium and the possibility of it being what we currently think of as Dark Energy.

Resolution of
Cosmological Issues – Ch.3

☒ Explanation of Work Energy
☒ Foundation of Quantum Mechanics
☐ Magnetic Monopole Mystery
☐ Missing Anti-Matter Mystery
☐ Neutrino Oscillation
☒ Origin of Cosmic Web
☐ Origin of Dark Energy
☐ Origin of Dark Matter
☐ Origin of EM field
☐ Origin of EM Waves
☒ Origin of Fields
☒ Origin of First Stars
☐ Origin of Force
☐ Origin of Gravity
☐ Origin of Mass
☐ Particle Decay
☐ Particle Evolution Sequence
☒ Particle Tunneling
☐ Purpose of Entropy
☒ SMBH Creation
☐ Space & Time Relativity
☐ The Fine-Tuning Problem
☐ The Flatness Problem
☐ The Hierarchy Problem
☐ The Horizon Problem
☐ The Magnetic Field
☐ The Measurement Problem
☐ The Reality Problem
☐ Virtual Particles
☒ Wave-Particle Duality Resolution
☐ What Drives Cosmic Evolution
☒ What is Energy
☐ Why Gravity so Weak
☐ Why Photons are Massless
☒ Why Three Generations

Complexity arises in nature not from a sudden and inexplicable appearance but by evolving in steps, with each having a demonstrated causal connection to a pre-existing state.

Chapter Four
4.0 Elementary Particles & the Intergalactic Medium

We are working under the hypothesis that the initial speck of Penergy expanded, thinned, and spun itself into SMBHs and particles, leaving a residual thin-density intergalactic Penergy medium (IPM). We are entering the perilous part of the story that hypothesizes new particles without a means to prove their existence. The motive driving this risky venture is that doing so provides answers to several important cosmological mysteries and completes the vision of how our Sub-A's *evolved* out of Penergy rather than having simply converted from a hypothetical *inflaton field*.

4.1 Early Elementary Particles

Those particles that spun directly out of the Penergy cannot be today's Sub-A's. According to the Standard Model, the first Sub-A's included electrons, photons, quarks, neutrinos, their extended generations, their anti-particles, and all the bosons. They were formed within the first second of the big bang and immediately began combining. That scenario requires far too many distinct particles popping into existence capable of complex interactions. Complexity arises in nature not from a sudden and inexplicable appearance but by evolving in steps, with each having a demonstrated causal connection to a pre-existing state. Penergy turning itself directly into those complex particles without a causal evolutionary foundation seems inconsistent with the way nature works.

We are therefore compelled to hypothesize that the freshly spun, simple, early elementary particles are new and different, requiring new names and descriptions. Our further development of the early universe remains as much as possible based on keen observations and good deductions, but we are diverging significantly from current theory. Two levels of particle matter below the subatomic level will be introduced. This picture of the early universe is unique but remains very plausible and relatively straight forward. It is a story about simple particles that combine and evolve into more complex particles. What follows is not meant to be an exact picture of how the earliest particles evolved but only how they could have evolved.

The first stable particles were truly elementary. It means that our electrons and quarks must be composite particles built from those elementals. Although there is no evidence for it, the argument for composite Sub-A's is reasonable and compelling. Consequently, we will proceed with the notion the early elementary particles are new to us and comprised of nothing more than tightly wound Penergy. By spinning into existence, these particles would presumably look like tiny tornadoes, so let's call them Torons, or just Tors for short. Everything we say about them will be new and not about any existing Model, so let's call all the related observations and deductions part of a *Tor Model*.

The creation of these early elementary particles ushered in new dimensions to our universe. The creation of distinct bits of particle matter with a finite distance between them implies *space*. Particles flying away from each other at a finite speed implies *time*. These particles were born into an ever-expanding cloud of Penergy; a thinning texture of the intergalactic medium we often refer to as *spacetime.* It is called spacetime because as it turns out the *space* and the *time* are relative to each other, but spacetime is still nothing more than our thin-density Penergy medium. Space and time are relative because time changes whenever space (*Penergy*) changes. A new theoretical basis for how space and time are relative is examined in Appendix #4.

We've learned from observations at particle accelerators and colliders that a high energy particle called a photon can decay into particles of matter. The decay results in a pair of particles, one spinning to the left and one to the right, which fly off in opposite directions. Particles formed through the decay of a large particle created in the Large Hadron Collider (LHC) often results in the creation of a particle pair. From those observations it is reasonable to conclude, as the Standard Model does, that the original creation of particles resulted in opposite-spinning particle pairs. Let's speculate on how that may be possible.

4.1a Elementary Particle Creation

At the very-light Penergy density stage at the end of the Cascade Effect the swirls would have been tiny and the spin start-up speed quick, creating particle-size matter. The sequence for elementary particle creation is illustrated in Figure 4.1 below.

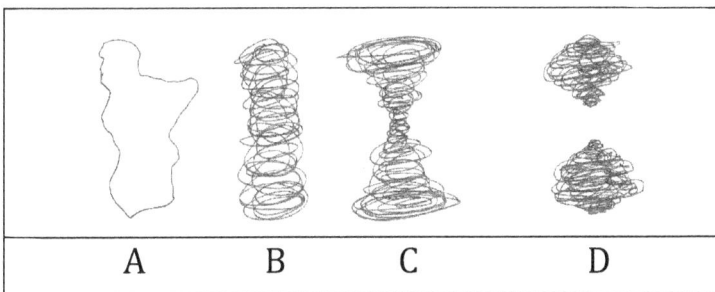

Fig. 4.1 The four stages of elementary particle creation.

67

(A): A fluctuation in the Penergy density causes a small glob to separate itself from the surrounding Penergy and begins swirling, like a tornado starting up. (B): As the spin rate goes up, the glob elongates. (C): As the spin rate goes up further, the elongated mass begins pulling apart, like a cell does at mitosis. (D): At near full spin-speed the elongated swirl divides, creating two distinct particles that speed away in opposite directions, one spinning clockwise and one counter-clockwise, conserving angular momentum. Once reaching maximum angular momentum, the particles shed excess mass back into the medium until the angular momentum energy matches the mass energy, giving them stability as calculated in Section 3.1. All of that could take place in less than a trillionth of a second.

4.1b Comparative Particle Attributes

The fast-spinning Tor particles would not result in loosely wound Kansas-like tornados. Being created directly out of Penergy they would be very tightly wound, giving them a definite size, location, mass, and a solid-like composition. Being tightly wound they would not be shaped like an elongated Kansas tornado, but more likely in the shape of a sphere, as we find spinning stars, planets, and blackholes. Created in pairs having opposite spins means they each have a spin axis and an orientation of top and bottom, one might refer to as a nose and a tail, respectively.

We don't know the specific attributes of these early elementary particles, but we might gain a picture of their possible attributes by making observations about the Sub-A's into which they evolved. The natural tendency for Penergy is to swirl and spin, which is how Tors came into existence. Consequently, we can surmise that Tors spin and do so on their own axis. As previously observed, they spin at a very high rate, at a calculated spin-rate vs mass ratio that gives them stability in the surrounding thin-density Penergy medium. Being massless and spinning at a very high rate, they move at light speed. And like massless photons, their defined travel speed is acquired by their relationship with the Penergy medium. (*How particles can be massless is examined in Appendix #1*)

Spin is important to the workings of Penergy. It is how it changes from a formless cloud into defined entities with mass, momentum, and other important characteristics. Spin is at the core of everything in existence, making it one of the most important dynamics in the universe.

We have observed our Sub-A's to be quite resilient. They can be moderately knocked about like billiard balls without losing their composure or identity. Since Sub-A's and Tors are made from the same state of Penergy, we can conclude that Tors are equally resilient, if not more so. Their resiliency may be the reason we cannot yet breakdown Sub-A's into deeper levels of constituent particles. The deeper we go, the smaller, more compact, and more resilient those particles are likely to be. We may never see early elementary particles and may forever know them only theoretically.

We have observed that Sub-A's are known to have *fields*. A field is an area of influence immediately surrounding the particle, as described in Section 1.1. We can surmise that the spinning gyrations of Tors disturbs the Penergy medium producing a personal field.

As mentioned earlier, when matter particles are created out of Penergy they are created in pairs with opposite spins. The pair of particles are labeled *matter* and *anti-matter*. The two particles have the same mass and are otherwise alike in every way except for their direction of spin. The particles are said to be a mirror image of each other.

When matter and anti-matter particles meet, they annihilate each other, turning back into a state of pure energy or photons. In the early particle creation era, there was a great deal of particle annihilation and photon production that created a very hot frenzy of activity. But, if matter and anti-matter were created in pairs simultaneously and therefore in equal numbers in the early universe, why is there predominately only matter in the universe today? What happened to the anti-matter?

4.2 The Mystery of the Missing Anti-Matter

No one knows for sure where the anti-matter went; it is one of the universe's biggest mysteries. The leading theory speculates that during the creation/annihilation period, out of every billion annihilations there was an imbalance leaving a single matter particle.[1] If this phenomenon happened over and over, it would leave the universe with an abundance of matter particles with few anti-matter particles, which is what we observe today.

That scenario is difficult to believe given the enormous number of matter particles in the universe and the low probability of the same imbalance happening over and over for all known particles. It's also a stretch to imagine the timing of it, since all the matter particle imbalances would have to have occurred within the first second of the big bang. That is a lot of annihilations, creations, and imbalances to balance out in such a short time. The annihilation imbalance theory stretches plausibility too far. There must be a better explanation.

There are other theories for the missing anti-matter, but the least complicated and most straight forward theory is that the anti-matter isn't actually missing. In the Tor Model, matter and anti-matter have opposite electric charge values[2], with matter being negative and anti-matter positive. The universe is said to be charge neutral, meaning equal in charge value.[3] For the universe to be charge neutral, the anti-matter must still exist. But if the anti-matter still exists, where is it?

4.2a The Case of the Anti-Missing Anti-Matter

The most straight forward answer may not seem feasible at first glance, but it develops into a very reasonable hypothesis. If the SubA's we know to exist – the quarks, electrons, and neutrinos – are composites of lighter elementary particles, then the missing anti-matter could be a part of the constituent particles that make up the Sub-A's. Anti-matter could be an integral part of the matter we see all around us, as will be demonstrated in the discussion on forces in Chapter Five. Assuming for the moment that notion is even possible, is there any observable evidence for the hidden presence of anti-matter?

Again, particles of matter and anti-matter are mirror images of each other, differing only in the direction of their spin and charge. Charge, the origin of which will be discussed in Chapter Five, can be either negative or positive; attributes ascribed to matter and anti-matter respectively. To find anti-matter therefore we only need to hunt down particles carrying a charge opposite that of matter. But we know of those particles already. Electrons have negative charge and protons have positive charge. Up quarks have positive charge and down quarks have negative charge.

Assuming for the moment matter and anti-matter carry opposite charges, and we find both types of charge existing in the particles around us, it suggests each type of charge and type of matter exists respectively within those particles. If the negatively charged electrons are referred to as matter, then positively charged protons should contain at least some anti-matter. These are bold ideas with some obstacles to overcome but given no better explanation for solving the anti-matter mystery, they are worth examining.

We must first consider our biggest obstacle: how does one construct composite particles by mixing matter and anti-matter together without them annihilating? As my grandmother would say, 'That problem's a doozy!'

4.2b Mirror-Image Annihilation

Annihilation occurs between colliding particles when the two are mirror images and attracted to each other, as are all matter/anti-matter pairs. An electron and its mirror image the anti-electron will annihilate, but an electron and an anti-muon (*a second-generation version of an anti-electron*) will not annihilate.[4.] Neither will an electron and antiproton.[5.] In each case the two particles are variations of matter and anti-matter, but the mass difference and perhaps other attributes between the respective particles precludes annihilation.

To annihilate, the two particles must be perfect mirror-images of each other, and a deviation in that perfection will preclude annihilation.[6.] The Tor Model posits that once constituent particles are bound together, they are no longer subject to individual particle annihilation. Consequently, a composite particle configured of three Tors would not annihilate with an opposite spinning composite of five Tors. The difference in the composite particle total mass would preclude annihilation.

In conclusion, we can reasonably conclude that only perfect mirror-image particles annihilate, and that particles with opposite spin and charge don't annihilate when their masses or other traits differ.

Experimental Prediction: Experiments demonstrating the successful mixing of bound states of matter and anti-matter, such as a lithium-matter nuclei and a helium-anti-matter nuclei, will confirm that only perfect mirror-image composite particles annihilate, paving the way for a complete, composite-particle theory.

With the notion that bound matter and anti-matter can exist within the same particle if they are not mirror images, we are set to proceed with the hypothesis that our Sub-A's are composite particles that may contain both matter and anti-matter. Such configurations will be examined in Chapters Seven and Eight.

Current particle theory denies the existence of particles being constituents to our Sub-A's, but many physicists don't reject the possibility. Physicist Jon Butterworth points out that what we think of as fundamental particles might simply be the smallest things we can measure with the tools available, and there may be deeper layers.[7.] Physicist Bruce Schumm tells us, "No one claims that quarks have been shown definitively to be the fundamental building blocks; it's just that with any experiment that we can do today, we can't 'see' anything smaller than roughly 10^{-18} meters across, and quarks are apparently smaller than that."[8.]

The idea of composite Sub-A's was momentarily popular in the 1970's. The elemental particles were called *Preons*. Several of the Preon theories were believed to be quite elegant and provided answers to outstanding questions within the Standard Model. Those theories, however, could not explain the unknown force holding the particles together and by the 1980s the theories lost their following.[9.]

The Tor Model can explain how elementals are held together, and the idea of composite Sub-A's solves too many outstanding mysteries to be ignored. Consequently, we will venture forward with the notion our Sub-A's are composite particles comprised of elements of both matter and anti-matter. First, we need to examine the next stage in the evolution of the falling Penergy density, when it became too thin for particles to even be created.

4.3 The Intergalactic Medium

The hypothesis here is that the final stage of the Penergy thinning process resulted in the creation of a pervasive thin-density Intergalactic Penergy Medium (IPM). Scientists have speculated on the origin and nature of a *dark energy* believed to be accelerating the expansion of the universe. There are a few theories for it, but once you get past the notion the initial speck of expanding Penergy was not consumed in the first second of the big bang, and that the speck is the only form of physical energy to exist in our universe, then it is quite feasible for the dark energy expanding the universe to be a residual thin-density Penergy medium.

Space is often referred to as the 'vacuum of space', but it is surely no vacuum. A physicist quoting Quantum Field Theory would describe it as a set of unseen, ubiquitous fields from which particles and everything else are made. An astrophysicist quoting General Relativity would call it 'spacetime' and refer to it as a *fabric*, which Astrophysicist Paul Sutter describes as a dynamic, living, breathing, physical object.[10.] An astronomer may refer to it as the Dark Energy that is expanding the universe. Someone more philosophically inclined might say it is all three: a physical, fluid-like, spacetime energy that is expanding the universe and can support fields. That description perfectly matches a thin-density version of Penergy.

As the universe expanded and Penergy thinned, we have surmised that it eventually reached a density that allowed SMBHs and particles to spin into existence. There is no reason to believe that all the existing Penergy was consumed in that SMBH-particle creation epoch, leaving a thin density Penergy medium. That prospect is not only plausible but seems obvious since, according to quantum theory, virtual particles can be momentarily created by borrowing energy from the vacuum of space. What other energy could those virtual particles borrow from if not the same energy from which all real particles are made? For virtual particles made of Penergy to pop into existence from the vacuum, that Penergy must be present in the vacuum.

Astronomer Mario Livio points out that particle-antiparticle pairs can also borrow energy from the vacuum, and those particles immediately annihilate themselves back into the vacuum. He characterizes the vacuum as bubbling with such virtual pairs.[11.] According to

Physicist Lisa Randall and others, when particles annihilate, they turn back into pure energy.[12] Again, if known particle pairs are being momentarily produced out of, and disappear into, the vacuum of space, that vacuum must contain the requisite energy to produce those pairs and be the reservoir for that disappearing pure energy.

Einstein assumed that space had a fluid-like nature, which proved to be a good approximation to the very large-scale structure of the universe.[13] Evidence that our intergalactic medium is a fluid-like substance comes from Einstein's General Relativity that predicted gravitational waves. Gravitational waves are disturbances in the intergalactic medium caused by cataclysmic events such as the crashing together of neutron stars or blackholes. Those disturbances create waves that propagate across space.

Gravitational waves stretch and distort along their path; a circle becomes an ellipse, and a square becomes a rectangle.[14] According to Professors Jorge Cham and Daniel Whiteson, this kind of behavior can only happen if space has a certain physical nature to it.[15] There must be something doing the waving and stretching. Referring to an intergalactic medium, Professor Wouter Schmitz says, "We just know that it must be there, and that it can carry a wave."[16] There seems to be little doubt that our intergalactic medium is real and has a fluid-like nature.

The presence of the Penergy may be difficult to prove. It is not a physical substance as we know it. It is not made from particles that would have mass or possess other particle-like characteristics that we can detect. Physicist Laura Mersini-Houghton tells us, "Dark energy behaves like an ether: its energy mysteriously diffuses out of a vacuum and yet it permeates every speck of the universe and pervades the very fabric of space-time."[17] Penergy, like an ether, may not be directly detectable, but we can infer its presence from its relationship to particles and virtual particles, as discussed above.

Einstein's theory of General Relativity taught us that gravity is the curvature of spacetime, which is often referred to as a fabric. That notion implies that space can be physically curved and distorted since that is the only reasonable way it could influence the path of passing matter. Physicist Paul Davies tells us that space is probably elastic. It can not only expand, stretch, and swell, it can buckle, twist, and shudder.[18.] Obviously, many believe that our spacetime vacuum is a real, physical thing with real traits.

All these characteristics make sense if they apply to the existence of a fluid-like medium. That medium may be difficult to detect because it is not like anything else we encounter, but there is little doubt the vacuum of space is a medium filled with some form of fluid-like energy. In keeping with our two fundamental principles, the energy that makes most sense is a thin-density version of the original Penergy from which the universe began. The presence of a medium of energy may mean that our concepts of spacetime, space vacuum, a cosmological constant, and dark energy are all the same thing manifested in the concept of an expanding intergalactic Penergy medium.

In conclusion, if all the Penergy was not consumed creating particles in the early universe, it would leave the residual Penergy growing ever thinner as the universe expanded. That Penergy density is apparently just below the particle-creation density threshold but is dense enough for fluctuations in that density to momentarily bring particles into existence. This remaining very-thin density of Penergy could now serve as our intergalactic medium and be what we perceive as the dark energy that has expanded our universe since its inception. It is the same Penergy from which everything is made and that serves as the reservoir from which virtual particles and pairs borrow, and into which they disappear.

4.4 Conclusions to Part I

The universe began as a speck of pure energy (*Penergy*) that began expanding and losing density. We know that particles are made from that energy, and a likely way for energy to become particles is for them to spin into existence. Particles demonstrate measurable wave-like characteristics, which means there must be a fluid-like medium in which those waves are produced. That medium is very plausibly the residual of the expanding cloud of thinning Penergy that permeates the universe and has been the cause of its expansion since birth.

Within the Penergy medium, particle waves are created by their spinning gyrations. A particle producing measurable waves must be spinning on an axis as opposed to having some sort of *intrinsic* spin. A particle spinning on an axis within a fluid-like medium will create a swirling of the medium producing a field surrounding the particle.

A particle spinning on its own axis that produces a field suggests the existence of a true fundamental particle like the Tor. If a Tor-type, fundamental particle exists, it means that our Sub-A's are composites of that particle. If our Sub-A's are composites, then anti-matter could be an integral part of those composites, solving the missing anti-matter mystery.

If the particles, the IPM, and personal fields are real, it lends credence to SMBHs spinning out of the medium-density Penergy in the early universe. The structure of the cosmic web is explained by the early development of pure-energy SMBHs and the Cascade Effect. The existence of SMBHs, particles, and the IPM means that Penergy is a physical energy manifested in different densities and is the ultimate source of fields and photons, the two sources of work energy.

All the Penergy has been consumed in the creation of SMBHs, particles, and the IPM, but its dynamic existence continues to play a significant role because the future evolution of the universe is founded on relationships between those three entities. The first relationship to evolve was between particles and the IPM that created fields, as relayed in Section 1.1. The next relationship to evolve was between the spinning fields that created *forces*, which will be examined in Chapter Five.

We have laid the foundation for our cosmic puzzle. The picture is credible, simple, and elegant. In Part II we lay the meatier puzzle segments in the way of forces and complex matter. These puzzle segments are a bit more speculative, not having quite the observational evidence for support, but they are equally plausible and will prove to be important puzzle pieces demonstrating the continuity of this new vision of cosmic evolution.

Resolution of
Cosmological Issues - Ch.4

☒ Explanation of Work Energy
☒ Foundation of Quantum Mechanics
☐ Magnetic Monopole Mystery
☒ Missing Anti-Matter Mystery
☐ Neutrino Oscillation
☒ Origin of Cosmic Web
☒ Origin of Dark Energy
☐ Origin of Dark Matter
☐ Origin of EM field
☐ Origin of EM Waves
☒ Origin of Fields
☒ Origin of First Stars
☐ Origin of Force
☐ Origin of Gravity
☐ Origin of Mass
☐ Particle Decay
☐ Particle Evolution Sequence
☒ Particle Tunneling
☐ Purpose of Entropy
☒ SMBH Creation
☐ Space & Time Relativity
☐ The Fine-Tuning Problem
☐ The Flatness Problem
☐ The Hierarchy Problem
☐ The Horizon Problem
☐ The Magnetic Field
☐ The Measurement Problem
☐ The Reality Problem
☐ Virtual Particles
☒ Wave-Particle Duality Resolution
☐ What Drives Cosmic Evolution
☒ What is Energy
☐ Why Gravity so Weak
☐ Why Photons are Massless
☒ Why Three Generations

Part II

The Heart of the Puzzle

Charged particles may absorb and emit photons, even virtual photons that may affect each other's momentum, but that fact by itself does not necessarily create an electric force.

Chapter Five
5.0 The Theoretical Basis for Force

In Part I we reframed our picture by restricting our initial conditions to plausible changes in the speck of expanding Penergy, and we built a solid foundation from puzzle pieces that seemed to naturally fit together. We now turn to the heart of the puzzle where forces act on matter to create complexity. Many large segments have been successfully completed showing the relationships between matter & energy, gravity & acceleration, photons & matter, electric & magnetic, and space & time, to mention a few. Before those segments can be brought together, however, we need to work on a core relationship that is essential to all of them - forces.

The first relationship to naturally evolve in our universe was between spinning particles and the IPM that created fields. We now examine the second relationship to evolve, which was between individual spinning fields that created forces. Force is an external agent that causes or changes the movement of a particle or mass. Like energy and fields, we cannot see forces, we can only witness the results of their presence.

The Standard Model posits particles interacting through one of four forces: electromagnetic, strong, weak, or gravitational. It posits those forces coming into existence within the first second of the big bang and being effectuated by the exchange of virtual particles called bosons.

Particle interaction is governed by certain rules that describe the four forces. According to Mathematician Milo Beckman, "The exact rules for particle interaction in the Standard Model are absurd.... The calculations involve continuum-sums and imaginary numbers and coupling constants and all sorts of ridiculous math..."[1] Physicist Michio Kaku tells us the Standard Model was created by splicing together by hand the theories that described the various forces, so the resulting theory is a patchwork. He reports that one physicist compared it to taping a platypus, an aardvark, and a whale together and declaring it to be nature's most elegant creature.[2]

Make no mistake, the Standard Model's equations describing particle interactions produce accurate predictions. But Dr. Beckman's comments and Dr. Kaku's imagery are good reminders the Standard Model often makes the math work while creating a questionable *interpretation* of the underlying phenomena.

The Tor Model does not allow anything to pop into existence and therefore sees force as something that *evolved* out of the conditions of the early universe. Let's examine the idea of force in each model, using the electric force as an example.

5.1 Force According to the Standard Model

In the Standard Model the electric force is conveyed by exchanging *virtual photons* between charged particles. A photon exchange between two like charges (*two protons*) creates a repulsive force, and between two unlike charges (*electron and proton*) an attractive force.

It is difficult to fathom how the exchange of a particle of any kind can create a precise attractive and repulsive force. The theory is that when an electron or quark emits a photon, it changes that particle's velocity, and when another particle absorbs the photon it changes that particle's velocity, just as if there had been a force between the two particles.[3.] It is especially difficult to understand how the exchange creates a consistent force when the exchange particle in question is a *virtual* particle. The following observations make the Standard Model's force theory by virtual particle exchange difficult to believe.

- A virtual particle is one that theoretically pops in and out of existence. It would seem impossible to have a consistent and smooth *force* while relying on a particle whose existence is fortuitous? This is especially mystifying considering each charged particle is theoretically exchanging an untold number of photons with every other charged particle at the same time. Virtual particles hypothetically do not possess a specific energy, so how can the impact of their absorption be relied upon to deliver a consistent quantity of force?

- Virtual particles are hypothetical, never directly observed. They are only believed to be real because there is no other reasonable explanation for such things as the Casmir Affect, and because mathematically if the virtual particle's presence is not included in the calculations the answers are wrong.[4.] So, the force by virtual-particle exchange theory relies on a hypothetical particle, with a hypothetical energy value, having a hypothetical effect on another particle. A reliance on so many extrapolated hypotheticals raises doubt as to the validity of such a scheme.

- Virtual-particle exchange often results in calculations that include *infinities* and the only way to make sense of the calculations is to simply ignore or remove them. Removing them has been given the innocuous name of *renormalization*; a procedure that is admittedly dubious mathematically.[5.] None the less, the interaction between photons and matter is summarized in a body of equations known as Quantum Electrodynamics (QED), which has proven very accurate and is used throughout science and industry. Again, the Standard Model has found a way to make the math work, but that doesn't mean the underlying theory involving the exchange of virtual bosons is correct.

- Of the four virtual-exchange particles, the Graviton, theorized to convey the gravitational force, has never been detected. The absence of the graviton brings doubt to the entire scheme of boson exchange being the source of force. Even if gravitons are found, the interaction between matter and gravitons produces infinities that cannot be renormalized, so at present graviton exchange is not even a viable theory.[6.]

- Physicist Kenneth Ford and others tell us that every time an electron absorbs or emits a photon, entire new particles are created, and the original particles are destroyed.[7.] If the original emitting electron is turned back into a knot of Penergy and from that knot a whole new electron and photon are created, and this happens every time a photon is absorbed and emitted by every charged particle, it seems like a *very* laborious, inefficient, and unreasonable way for nature to work.

- The Tor Model posits force being conveyed through the fields of particles. Scientists have noted that fields take time to exert their influence. Physicist George Musser notes that a time lag seems odd if forces are leaping directly from one object to another [*as in particle exchange*] but a time lag is perfectly natural if an impulse must make its way through a medium.[8.] In other words, one would expect a force conveyed through a boson exchange to be initiated smoothly, without a time delay, but evidently that is not what scientists observe. The noted time delay makes sense if forces are conveyed through fields in a medium, as they are in the Tor Model.

- Two charged particles could theoretically feel a force between them even at galactic distances. According to Physicist Lawrence Krauss, an electron on earth could theoretically have an influence on an electron on Alpha Centauri, four light years away.[9.] It would mean that an electron could be constantly exchanging virtual photons with any of the many charged particles in the space in between! Again, that makes this scheme laborious and inefficient, making it difficult to believe nature works that way.

- Theoretically, for virtual particles to come into existence they borrow energy from what is called the vacuum of space. If virtual bosons are popping in and out of existence constantly to affect forces, it would require a great deal of energy to bring all of them constantly in and out of existence, even momentarily. Quantum theory can predict how much energy density there should be in the vacuum of space by adding all the virtual particles together. Doing so produces a theoretical energy value of 10^{105} Joules per cubic centimeter. Examining space, however, produces a value of only 10^{-15} Joules per cubic centimeter. Comparing what the energy level theoretically should be to what is measured, scientists find the theoretical value to be 10^{120} times larger than the actual value.[10.] Michio Kaku characterizes this as the largest mismatch in the entire history of science.[11.] Virtual particles might exist, but apparently not even close in numbers to support them being the source of all the forces.

According to quantum physics, the vacuum of space is awash in virtual particles constantly popping in and out of existence. The Tor Model recognizes the possible existence of those virtual particles but not to the degree of existence theorized by quantum physics. In the Tor Model those particles are created only when a fluctuation in the Penergy medium creates a small knot that provides a momentary rise in the Penergy density sufficient to bring particles into existence. But because the knot begins immediately dissipating back into the surrounding thin-density Penergy medium, the failing knot is not strong enough to sustain itself, so the particle(s) disappear back into the medium before being fully formed.

Consequently, virtual particles, if they exist, are not popping in and out of existence everywhere, only where fluctuations in the Penergy medium are prominent such as near blackholes and other very dense masses, or within the environment of a planet's atmosphere such as our own. Charged particles may absorb and emit photons, even virtual photons that may affect each other's momentum, but that fact by itself does not necessarily create an electric force.

We will now examine force from the perspective of the Tor Model as being a relationship between personal fields, and we will examine how that works in each of the forces.

5.2 Force According to the Tor Model

A force effects all particle interactions resulting in one of three relationships: (1) The force can cause particles to be so attracted to each other they are said to combine; (2) The particles can be attracted to each other but unable to combine; and, (3) There is no attraction and the particles are either neutral or repulse each other. All these interactions can be understood by examining the relationships between particle personal fields. To understand how fields produce a force, let's examine the force that allowed the first particles to combine - the magnetic force.

5.2a The Magnetic Force

Scientists have observed that when magnetic attraction comes into play the particles involved have their spin axes lined up in the same direction. A natural magnet called lodestone (*magnetite*) is magnetized by virtue of the electrons in its atoms having their spin axes aligned. When the lodestone is placed near a metal object, the magnetic field around the lodestone encompasses the metal object causing the spin axes of the metal object's electrons to align in the same parallel direction. That effect on the adjacent atoms that cause them to align is said to be a *force* applied by the magnetic field.

The magnetic body with its electrons aligned is said to have a head and a tail, or a north pole and a south pole. When the north pole of one magnet is closely aligned with the south pole of another magnetized object, they are attracted to each other. But when they are aligned otherwise, especially north to north or south to south, they repulse each other, as you know if you've ever played with magnets. But what is causing this unseen force that aligns spin axes and attracts magnets?

The Tor Model sees the source of the magnetic force as the relationship between the personal fields of like-spinning early elementary particles we call Tors. The Tor particle spin causes the surrounding Penergy medium to spin with it and fan out creating a personal field. When the spinning field of one Tor comes into a close nose-to-tail alignment with the like-spinning field of another Tor, the fields are attracted to each other and will be drawn close together with a strength consistent with their natural coupling capacity. Observational evidence for the like-spinning fields being aligned may lie in the fact bar magnets show a spinning vortex at the end of each pole. If the spin axes alignment between fields is otherwise, the fields repulse each other. This is the basis for the magnetic force, and it is due solely to the spin dynamics and fields of the respective like-spinning Tor particles.

The magnetic field is represented by iron filings on a paper placed above a bar magnet, as shown in Figure 5.1. The magnetic field lines loop through the north and south poles, entering south and exiting north. When two bars are placed in line nose to tail, those field lines continue through both bars creating a natural attraction between the north and south poles. We also observe that the strength of the field is greatest near the poles and falls off as one moves away from those poles, just as the particle's personal field would do.

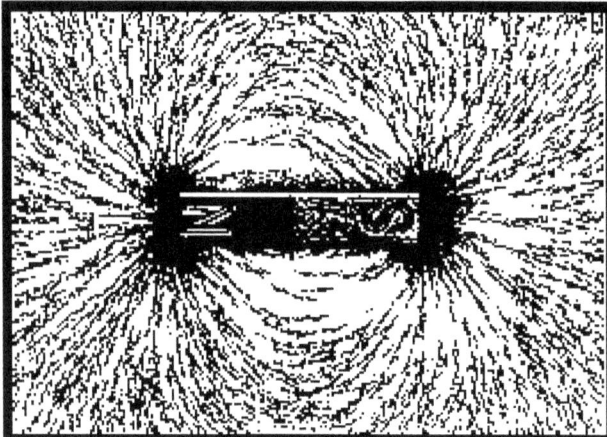

Figure 5.1 Iron filings on a sheet of paper covering a bar magnet showing its magnetic field.

This picture of iron filings is a good example of particles producing a field around themselves and the cumulative effect of the overlapping fields spreading out and influencing other particles. In this case the aligned atoms (*and constituent Tors*) within the magnet are producing a magnetic field that is in turn influencing the atoms in the iron filings, causing them to move into a fixed orientation consistent with the magnet's field.

In summary, the spinning of a Tor particle creates a personal field and the effect that field has on the fields of other like-spinning Tors is called the magnetic force. The two like-spinning Tor fields will cause the fields to attract each other and join if closely aligned nose to tail, and repulse each other if meeting in any other arrangement.

5.2b The Creation of Tor-Chains

The magnetic force generated through Tor field attraction is what brings Tor particles together. When aligned nose to tail they are attracted with sufficient strength to link each Tor in a semi-fused relationship. Such in-line attachments would create a *chain* of like-spinning Tors, as illustrated in Figure 5.2.

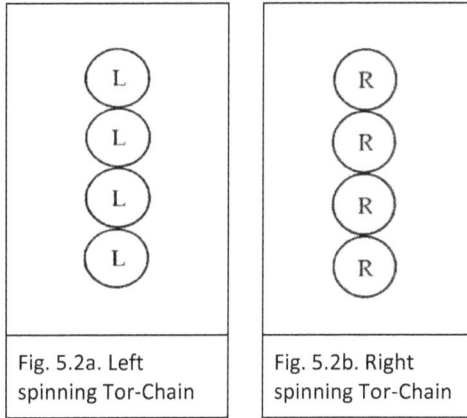

Fig. 5.2a. Left spinning Tor-Chain

Fig. 5.2b. Right spinning Tor-Chain

Figure 5.2 The strong nose-to-tail coupling between like-spinning Tor fields allows Tors to combine into Tor-Chains.

Tor-Chains can be left-spinning or right-spinning and can be any number of Tor particles long. Longer chains would make it less likely the chain would be annihilated by bumping into a mirror-image opposite spinning-chain of the same length. The longer the chain, however, the more likely it would be broken apart by a collision with a more energetic Tor-Chain.

A spinning Tor-chain creates a magnetic field that carries over to electrons and atoms. Consequently, bound particles such as electrons can have their spin axes aligned, and the alignment of multiple-particle spin-axes adds to the magnetic field. The more particles involved in the alignment process, the greater the size and strength of the magnetic field.

5.2c The Electric Force

Eighteenth century scientists and others including Ben Franklin were familiar with the rudiments of electricity being the motion of electric charges which Franklin labeled positive and negative. French scientist Charles-Augustin de Coulomb established in the late 1700's that charged objects attract and repel each other. The connection between electricity and magnetism was discovered in the early 1800s by Michael Faraday. Faraday formulated the notion of magnetic and electric fields, and that particles passing through those fields would experience a magnetic and electric force capable of changing the particle's velocity.

The electric force is characterized by the relationship between traits of sames and opposites. The Standard Model ascribes those traits as being positive and negative *charge,* an undefined characteristic inherent in the particle. The Tor Model ascribes those traits as being the Tor particle's field spin direction. The respective fields of the Tors create an attractive/repulsive electric force, with opposite spins being attractive and like spins being repulsive, except when the fields are aligned nose to tail.

Mirror-imaged elementary particles are opposites subject to an attractive/repulsive force. Particles identified as matter and anti-matter are opposites subject to an attractive/repulsive force. Particles said to possess positive and negative charges are opposites subject to an attractive/repulsive force. In all these cases the source of the attractive/repulsive force is the same – the electric force created by the relationship between the personal fields of Tor particles.

Given the arguments against the source of electric force being virtual boson exchange; and given the simplicity of that source being the natural attraction/repulsion between the fields of Tor particles; we can hypothesize that the source of the electric force is the relationship between particle fields as measured by a differential in the number of positive and negative Tors making up a particle. An abundance of positive Tors would give the particle a positive charge, and vice versa.

5.2d Rules for How Fields Affect Each Other

For clarity I will summarize the rules for particle field interaction. The fields of two of the same elementary particles spinning in the same direction are Highly Compatible, meaning that when the fields align nose to tail, they are attracted to each other and can join. When those fields are only closely aligned nose to tail there is enough attraction to bring them inline before joining. If the alignment is off significantly, the like-spinning fields are Incompatible and will repulse each other. These relationships are demonstrated in the attraction and repulsion of magnets.

The fields of two of the same elementary particles spinning in opposite directions are Somewhat Compatible, meaning that the fields of the two particles are very much alike, which makes them attracted to each other, but not being matched fields, they cannot touch or otherwise join. That relationship is demonstrated in the attraction of particles with opposite charge.

That is not the case, however, for complex, multi-particle composites. Two like-spinning (or like-charge) composites aligned nose to tail may have a mild attraction, but they will not join, except under unusual conditions, as when all of their energy has been removed, leaving them barely able to move. Two electrons in that state are called Cooper Pairs. Perhaps taking their energy affects their spin rate, that in turn weakens their field, limiting the capacity to repulse each other.

In conclusion, what we know as positive verses negative charges, and matter verses anti-matter, is simply the relationship between left and right spinning Tor fields. The respective fields create an attractive/repulsive electric force, with opposite spins being attractive and like spins being repulsive, except when the fields are aligned nose to tail.

5.2e Particle Assembly from Tor-Chains

Since mirror-image Tor particles have a natural attraction for each other, it is reasonable to assume that mirror-image *Tor-Chains* would likewise have a natural attraction. Exact length, mirror-image Tor-Chains would annihilate, but if they had different lengths, they would attract each other without annihilation.

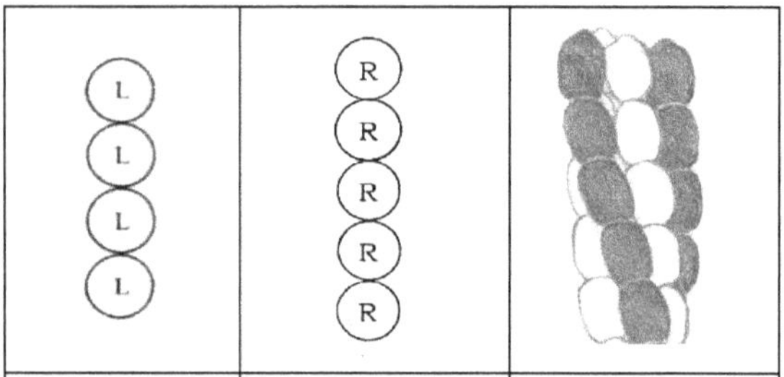

Fig. 5.3a Four particle, Left Tor-Chain.	Fig. 5.3b Five particle, Right Tor-Chain.	Fig. 5.3c Four Tor-Chains of different lengths.

Figure 5.3 Left and Right Tor-Chains of different lengths could combine without annihilation.

The attraction allows opposite-spinning Tor-Chains of different lengths to successfully combine, as illustrated in Figure 5.3c. Tor-Chain attraction could create combinations of various chain lengths with potentially any number of chains involved.

With aligned, like-spinning Tor fields creating a *magnetic force*, and opposite-spinning Tor fields creating an *electric force*, we have the forces necessary to construct composite particles.

5.2f The Creation of Tryks

Individual mirror-image Tors will annihilate, but when bound up in a group such as a chain, the individual Tors are no longer subject to annihilation. Only a mirror-image chain of the same length can annihilate the chain. The environment of many different chain lengths would have slowed the early cosmic annihilation rate. This would have allowed the chains to survive long enough to combine and become a part of a larger composite.

Having Tor-Chains and single Tors available to combine, the next stable combination may have been a long Tor-Chain encircled by any number of single Tors. For simplicity, that notion will be illustrated in Figure 5.4 as a four Tor-Chain encircled by a single opposite-spinning Tor. All kinds of different combinations are of course possible, and this configuration is set forth only as a simple working example. We will call this individual-Tor and Tor-Chain combination a **Tryk**. The mirror-image form of the Tryk configuration would also evolve and be stable, and as usual if the two should meet they would annihilate each other.

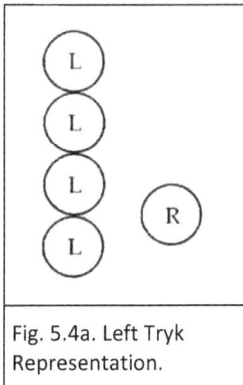 Fig. 5.4a. Left Tryk Representation.	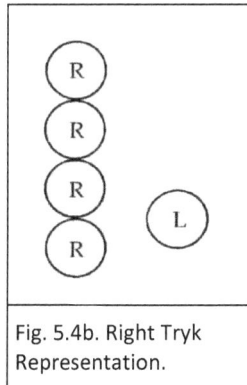 Fig. 5.4b. Right Tryk Representation.

Figure 5.4. Left and Right Tryks made from a Tor-Chain encircled by a single opposite-spinning Tor.

With a Tryk, the single opposite-spinning Tor in relation to the Tor-Chain is subject to the electric force and has the same rotational dynamics as that of an electron circling a nucleus to create an atom. Again, because their respective fields are somewhat compatible, there is an attraction, but because the fields are different, they will not touch and will repulse each other if they get too close. Given that Tors can combine into chains, and opposite spinning Tors and Tor-Chains can attract each other without annihilation, it is possible to have all sorts of combinations evolve.

Electric charge is the term used to ascribe an attractive/repulsive characteristic to a particle. That characteristic is simply a differential in left and right-spinning Tors. The imbalance is reflected in the particle's field. If a Tryk had a dominance of left-spinning Tors (*shown in Fig. 5.4a*) we could call it a positive charge field, and a dominance in right-spinning Tors (*shown in Fig. 5.4b*) a negative charge field. The strength of the electric charge is in proportion to the size of that Tor-spin count differential - the greater the Tor count differential, the greater the charge value. We will discuss particle charge values in detail in Chapter Seven.

While charge depends on the spin direction of its Tor particles, it does not apply to composites that can spin in either direction without regard to its charge.

5.2g The Electromagnetic (EM) Force

Since Maxwell brought the magnetic and electric forces together under a single set of equations, the two forces have been thought to be two sides of the same coin. The electric and magnetic fields are distinct, but subatomic particles show evidence of possessing both. This makes sense in the Tor Model, since all particles are made from combinations of Tor Chains and Tryks, which display magnetic and electric fields respectively. This gives particles both a magnetic and electric component and a field combination called the electromagnetic field that exerts an electromagnetic (EM) force. The EM force is normally measured by the strength of the particle's electric field.

The EM force is the relationship between the fields of electrically charge objects, with like-charges repulsive and unlike attractive. The strength of the EM field is proportional to the distance between charged particles. The closer one brings the charges (*fields*) together, the stronger the force. The combination of the incompatibility of the respective forces between particles, in combination with their kinetic energy, prevents particles from touching. In the Tor Model this phenomenon is believed to have its limits. When the fields get too close, the fields interfere with each other and begin losing their attractive/repulsive capacity. If the particles were compressed to the point they were forced to touch, their fields could interfere to the extent they would presumably begin to lose much if not all their attractive/repulsive capacity.

Experimental Prediction: Someday an enterprising physicist will create an experiment that will demonstrate that elementary particles when compressed together to the point of touching will lose much if not all their attractive/repulsive capacity, thus strongly suggesting that attractive/repulsive force is derived from particle personal fields, not the exchange of virtual bosons.

Charged objects are naturally surrounded by their personal field containing all their attributes, including the effects of their charge differential. If one lines up a row(s) of positively charged objects on one side and negatively charged objects opposite them, with their respective fields overlapping, it creates an attractive electric differential that will induce electrons to move across the combined fields creating electricity.

Given a reasonable distance between them, the more charged objects that are involved, the stronger the combined field. The strength of the electric differential between the positively and negatively charged rows is measured in *volts*. An electric charge differential can also be induced in a wire by the movement of an adjacent magnetic field, which is how a generator works.

In conclusion, the fields of like-spinning Tors are attracted to each other because the fields are compatible when aligned nose to tail, but incompatible and repulsive when aligned otherwise. With particles of opposite spin, the two fields are similar in nature but not exact matches. When two unlike charges approach each other the magnetic aspect causes their fields to orient in a parallel configuration making nose to nose, and tail to tail configurations seemingly compatible and thus attractive. But being different types of spin, the fields are not compatible to the extent of allowing a union, so the fields are attractive and can get close but not touch.

The Force (F) between two charged particles falls off inversely to the square of the distance between the two charges (q). The equation expressing this notion is, $F = kqq/r^2$. This point is important to remember when we discuss the gravitational fields in Appendix#3.

5.2h Monopoles

Scientists have long speculated on the possible existence of magnetic monopoles. In the Tor Model, since the magnetic field emanates from one or more like-spinning Tor particles that spin on their own axis, any combination creates a chain that is dipole by design. There are no magnetic monopoles. The only way to create such a particle would be for a Tor-Chain to link its nose and tail together in a loop, but that is a composite particle, not a single particle with a magnetic pole, and its field would be neither attractive nor repulsive.

It is different for the electromagnetic field. The creation of the electromagnetic field arises from a combination of left and right-spinning Tors that are together in the same particle, but they cannot line up in a single chain. Consequently, no matter how configured, the resultant particle is comprised of multiple spin orientations, so the particle cannot spin on its own single axis. The field created by the left and right-spinning Tors with different orientations cannot be in a dipole configuration and therefore appears as a monopole. The charge related to the monopole, of course, depends on the distribution of Tors; an excess right-spinning Tors creates a negative charged monopole, and an excess of left-spinning Tors a positive monopole.

5.2i The Gravitational Force

According to the Standard Model, Gravitons are the bosons mediating the gravitational force. That means that all the particles in your body are constantly exchanging gravitons with the earth, the earth is constantly exchanging gravitons with the sun, and the sun is constantly exchanging gravitons with the SMBH at the heart of our galaxy. It's seemingly the only way all those gravitational forces could exist if gravity worked by exchanging gravitons. That is a lot of long-distance, continuous particle exchanges holding our galaxy and solar system together, making it difficult to believe nature works that way. The graviton has never been detected.

Einstein gave us an alternative explanation by showing us that gravity is a curvature of the fluid-like nature of space, bending the pathway of nearby matter relative to the size of the respective masses. The Tor Model agrees with Einstein's General Relativity equations telling us how mass bends spacetime, but the equations don't give us a clear explanation of what gravity *is*. I've often wondered, what is inherent in mass that causes the surrounding medium to bend? My theory for gravity is rather speculative, so it is examined in Appendix #3, but the bottom line is that a gravitational field is created by the spin of Tors within the mass causing a compression of the surrounding Penergy medium toward the center of mass.

5.2j The Strong Force According the Standard Model

The strong force is said to exist within the nucleus, holding together the components making up the nucleons. The components and forces within the nucleus are largely hypothetical, so I will address the theories according to the Standard Model and Tor Model separately.

According to the Standard Model, the three quarks of a nucleon are held together by a strong force generated by the constant absorption/emission of virtual gluons. Gluons have not been seen directly but are believed to have come into existence momentarily in the Large Hadron Collider based by the analysis of high-energy particle collision debris. The strong force is unique in that the three quarks cannot be separated no matter how much energy is put into the effort. It is as if they are held together by strong rubber bands.

Quarks and gluons are theorized to possess a special kind of charge called Color Charge, which controls how quarks exchange gluons. A color charge has never been detected and is derived solely from a brilliant scheme developed by physicist Murray Gell-Mann in the 1970's that explains very accurately how this force might work. The Strong Force operates at very short distances and only inside the nucleus. The theory and math supporting the strong force are quite complex, there being three different color charges and eight different gluons involved.

5.2k The Strong Force According to Tor Model

It remains difficult to believe that a force is effectuated by exchanging virtual particles. There is also the argument that every time there is an absorption or emission, all particles involved are destroyed and new particles produced, creating the height of inefficiency. Consequently, it is worth exploring an alternative theory for the source of the strong force.

Scientists now believe a nucleon contains far more matter particles than just the three quarks defining its electric charge value. The mass of a nucleon is perhaps a hundred times the mass of the three defining quarks combined, so it is reasonable to surmise that there may be more to the nucleon than just three quarks.

To investigate the interior of the proton, scientists shoot high energy particles at it and observe how they scatter. The more energy the shooting particle possesses, the smaller its wavelength and the better definition the scattering theoretically provides.

In a 2021 Forbes Magazine article by Astrophysicist Ethan Siegel, entitled *What Rules the Proton: Quarks or Gluons?* he details what scientists find from those scatterings. He reports that low energy collisions are dominated by quark-quark interactions. Higher energy collisions start to see quark-gluon interactions, with some quarks turning into heavier charm quarks. Still higher energies are dominated by gluon-gluon interactions. He summarizes this by noting that at low energies a proton is more 'quarky', but at higher energies it's rather 'gluey'.

Within the Tor Model, this method of experimentally exploring the interior of a proton raises concerns. Perhaps shooting high energy particles into a proton doesn't necessarily give us a better definition of its innards but reveals the type of particles that could be *created* from the knots of Penergy produced by those high energy collisions. If the 'gluey' gluon state only shows up under high energy conditions, it suggests that gluons are not fundamental but only exist when there are knots of high density Penergy to produce them. The high energy collisions that create knots of Penergy are then decaying into smaller particles producing jets and the outline of particles that could be interpreted as being specific particles present inside the proton. These experiments might be telling us exactly what the scientists say they do, but given how easily knots of Penergy can create particles we must be careful in our analysis of such experiments. Shooting high energy photons into a proton could be creating particles rather than merely identifying existing particles.

The only force the Tor Model recognizes acting like a rubber band is the magnetic force working at the Tor level. As observed earlier, the creation of Tor-Chains is due to the powerful coupling capacity of the aligned fields of like-spinning Tors. Such a force might act like a rubber band if one tried to pull a Tor-Chain apart. This would be especially true if the connections were stranded into multiple Tor-Chains, as illustrated in Figure 5.3c.

This notion makes sense once we examine the possible spin dynamics of the Tors. Imagine the spinning Tors affecting the surrounding Penergy medium by causing it to rapidly swirl in the direction of the Tor spin. As the nose of one Tor field comes close to the tail of another Tor field, the swirls would align themselves and the natural coupling attraction would pull them together, uniting the two particles through their fields. The two fields would pull into each other without the two particles ever touching. As the particles approach each other the strength of the attractive coupling would relax and perhaps even become repulsive if the two particles moved too close to each other.

The two particles would remain in this relaxed, balanced state until some outside force affected them. If an outside force tried to pull them apart, the strong coupling capacity of the fields would again come into play resisting any separation as they were drawn apart. A force sufficient to overcome the very strong coupling capacity could separate the two fields, but as observed in particle experiments, when enough energy is put into the effort to separate the quarks, the knot of infused Penergy creates new particles that fill in the gap before the quark fields can be separated.

The very strong coupling capacity only occurs at the Tor level. It is a weaker dynamic at the atomic level. The aligned Tors create a magnetic field that affects the electron alignment in atoms making up bar magnets. While the aligned electrons create a magnetic force effectuated through the metal bars of the magnets, the strength of the attraction between the aligned electrons isn't near that of the underlying Tors. That is why it is relatively easy to separate bar magnets.

This scenario is a good argument for the quarks being connected by Tor Chains but there is no evidence for it. A complete theory may be better developed once the additional internal parts of the nucleon, if there are any, have been better defined. We must await further experimental developments before a complete theory can be developed.

5.21 The Weak Force

A stand-alone neutron is unstable and will decay into a proton, electron, and neutrino within a few minutes. Mid-twentieth century scientists were familiar with the magnetic, electric, and gravitational forces, so to explain how a neutron could decay into a proton, scientists had to define one more force.[12] A new particle had to be introduced to explain this force.

Large particles were predicted and ultimately detected in the Large Hadron Collider, dubbed the W and Z particles. They have not been observed directly but are believed to have been identified by interpreting high-energy particle collision debris.[13] According to the Standard Model these W and Z virtual bosons are the force carriers of the Weak Force that is responsible for the decay of the neutron and other particle interactions inside the nucleus. The weak force might be everything scientists say, but there are aspects of the theory that make it difficult to believe.

The W and Z particles are massive, being more than 80,000 MeV, compared to the light-weight quark at less than 5 MeV. The W and Z particles exist only briefly before decaying. They are only seen in the Large Hadron Collider under very high energy conditions, yet they are theorized to come into existence every time there is a decay/creation or absorption/emission within the nucleus in moderate energy conditions. Physicists have created the mathematics to support the notion these very massive particles come into existence by momentarily borrowing the requisite energy from the vacuum of space. As noted earlier, apparently the available energy for all those borrowings may not actually be present in the vacuum of space.

These massive particles may momentarily exist in colliders, but they might simply be first or second-generation creations that could not continue to exist in the thinning Penergy density as described in Section 3.4. They may have gone the way of all other first and second-generation particles, making their relationship to today's particles questionable. Let's examine particle decay from another perspective.

5.2m Particle Decay According to the Tor Model

In the Tor Model, particles decay (*decompose*) in two ways, *forced decay* and *natural decay*. Forced decay occurs when a particle cannot subsist in the existing density level of the Penergy medium. This occurs in an accelerator or collider when a knot of Penergy creates a heavy particle. The heavy particle immediately decays into lighter particles, and in some cases those lighter particles resolve into even lighter particles, until all particles produced can reside in the existing Penergy density.

Natural decay occurs when a large atom such as plutonium changes by ejecting an alpha particle from its nucleus, which is a process known as Alpha Decay. The cause of the particle instability causing decay is not well understood. In the Tor Model, Alpha Decay could be caused by the jiggling of a large nucleus that occasionally distorts in shape allowing the repulsive forces between protons to overcome the short-ranged nuclear force, enabling the alpha particle to escape. Chemical Physicist Michael Munowitz reminds us that despite the strong nuclear force, the electric repulsive force between protons is always present and the potential for disaster never disappears. When the repulsive force overtakes the strong force, radioactive decay results. The nucleus can spew out a proton, neutron, or alpha particle. After the decay, all the protons and neutrons are accounted for. The particles are shuffled about but are otherwise unchanged.[14.]

Natural decay also occurs when a down quark decays by ejecting an electron and neutrino, which is known as Beta Decay. In Beta Decay, like most decays, the particles are ejected at speeds approaching that of light.[15.] This suggests that the source of the sudden burst of energy triggering the decay might be from the absorption of a high energy photon that overcomes the binding energy of the electron connection, breaking that connection and allowing the electron-neutrino duo to escape from the down quark, producing an up quark.

Standard Model physicists would argue that the created particles from the beta decay did not come directly from the original particle, but from the virtual boson that mediated the decay. In the case of the decay of a down quark, the boson in question would be a virtual W⁻ particle. It means that the less than 5 MeV sized down quark suddenly emitted an 80,000 MeV virtual W⁻ that then immediately changed back into a less than 5 MeV up quark, plus an electron and neutrino. Assuming that is even possible, it seems like a laborious way for nature to work.

Heavy particles like the W boson are believed to exist and physicists have developed the math to support such an inefficient interaction, but relying on heavy, virtual particles to explain particle decay seems dubious. It is also unnecessary if our Sub-A's are composite particles. An equally plausible explanation for the result of the down-quark decay is again that the electron and neutrino were present within the parent down quark before the decay. Admittedly, there is no direct evidence for such a composite configuration, but we just don't yet have the technology to break down those tiny Sub-A's to that degree. An example of such a configuration will be provided in Chapter Seven.

As detailed earlier, physicists theorize that after each interaction, which includes decay, particles are not simply reconfigured but instead all new particles are created. This seems rather extreme, and it makes sense for nature to handle quark decay like it does atom decay. When an atom decays by spitting out an alpha particle, it simply leaves behind an atom with different atomic numbers and characteristics. The protons and neutrons are reshuffled but otherwise remain unchanged. Particle decay seems more about rearranging existing components than creating all new components from scratch.

Again, the weak force may be what physicists believe it to be, but under the Tor Model its existence for the purpose of particle decay does not seem necessary. Given the doubts as to the source of force being boson exchange, it's possible that the arguments for its need and existence can be answered by other means.

5.2n Conclusions

In conclusion, using the electric force as an example, we examined the feasibility of force being effectuated through the exchange of virtual bosons and concluded that the observational evidence and related rational conclusions bring that theory into serious doubt. We next examined the Tor Model's theory for force being generated through the relationship between personal fields, and we examined each field to see how that would work. From these examinations, we concluded that the attractive/repulsive characteristics involving matter/anti-matter, electric charge, and magnets are all derived from the same source – the attractive/repulsive relationship between the underlying fields.

Working Prediction: Once personal fields are proven to exist, a physicist or astrophysicist will extrapolate the mathematics of quantum field theory into a theory of the magnetic and electric forces conveyed through the personal fields of early elementary particles.

Our assembly of puzzle pieces reflecting the evolution of the early universe has been quite a journey. The puzzle is filled with maturing supermassive blackholes, cascade remnants of faltering smaller swirls, Tor particles, and a very thin Intergalactic Penergy Medium (*IPM*). The Tor fields, endowed with what we now call the magnetic force, allowed Tors to combine nose to tail into Tor-Chains. Through the electric force, Tor-chains could then combine with single Tors to create a composite particle we are calling a Tryk. With these evolutionary advances, we have the tools to move forward with the creation of puzzle pieces representing composite particles. But first let's take a look at the cosmic environment - the stage upon which all of this evolution takes place is an important aspect of the picture puzzle.

Resolution of Cosmological Issues - Ch. 5

☒ Explanation of Work Energy
☒ Foundation of Quantum Mechanics
☒ Magnetic Monopole Mystery
☒ Missing Anti-Matter Mystery
☐ Neutrino Oscillation
☒ Origin of Cosmic Web
☒ Origin of Dark Energy
☐ Origin of Dark Matter
☒ Origin of EM field
☐ Origin of EM Waves
☒ Origin of Fields
☒ Origin of First Stars
☒ Origin of Force
☐ Origin of Gravity
☐ Origin of Mass
☒ Particle Decay
☐ Particle Evolution Sequence
☒ Particle Tunneling
☐ Purpose of Entropy
☒ SMBH Creation
☐ Space & Time Relativity
☐ The Fine-Tuning Problem
☐ The Flatness Problem
☐ The Hierarchy Problem
☐ The Horizon Problem
☒ The Magnetic Field
☐ The Measurement Problem
☐ The Reality Problem
☒ Virtual Particles
☒ Wave-Particle Duality Resolution
☐ What Drives Cosmic Evolution
☒ What is Energy
☐ Why Gravity so Weak
☐ Why Photons are Massless
☒ Why Three Generations

For complexity to grow effectively, death and decay must be a part of the process.

Chapter Six
6.0 The Early Cosmic Environment

The continuous creation and destruction of Tor and Tryk combinations would constitute the evolutionary phase of the next building blocks. Many combinations would come into existence, be broken apart, recombine into different configurations, and be broken apart again. It's the same product strengthening process every evolutionary Level goes through before a stable, dominant form eventually evolves from the frenetic environment. This evolutionary process – the gradual development from simple to complex shaped by component sustainability - is called *natural selection*. It is one of many processes constantly taking place throughout the cosmic environment.

6.1 Survival by Natural Selection

The actual combination of Tors and Tor-Chains comprising the configuration of Tryks is unknown; they may be simple or quite complex. With all the collisions, annihilations, and particle-pair creations taking place, the Penergy medium would have heated up, becoming a hot, high-energy *frenzy* of activity. Only the strongest composite combinations would survive the frenzy. In this early stage of cosmic evolution, survival by natural selection had begun.

Biological natural selection is the process of organisms undergoing physical change due to DNA mutations. Some mutations make the organism better adapted to its environment, which raises its chances to survive and reproduce, perpetuating those survival traits. In the case of elementary particles, the strongest and most suitable composite combinations would survive the frenzied environment and go on to become stable and available to combine again, perpetuating its configuration. Collisions breaking up weaker connections that reconfigure themselves is the mutation equivalent for particles.

Tors and Tryks represent the first and second Level of matter, respectively. The third Level of matter, Sub-A's, will emerge from the frenzy of the environment to be quite complex compared to the first two Levels. For the third Level to evolve, many kinds of Tryk combinations will be made and torn apart, with their fragments recombining into new configurations. Our Penergy medium will have become a sea of Tors, Chains, Tryks, and complex combinations, much like the earth's oceans were a sea of macromolecules prior to cells evolving. All this particle creation, annihilation, and smashing combinations apart creates a Penergy environment that is truly a *frenzy.* Which combinations would survive annihilation or be broken apart by collision is impossible to say, but natural selection in this hot, frenzied environment would definitely determine their number and composition.

6.2 Evolution Progresses in Levels

There are many theories for evolution, each having some validity depending on their perspective. One is to think beyond planet earth; to step back and observe how matter has evolved cosmically over the entire history of the universe. Going back to the evolutionary level we recognize as the creation of subatomic particles - **Sub-A's**, it is theorized that quarks *combined* to form protons and neutrons. Then the protons and neutrons *combined* to form nuclei. The nuclei eventually *combined* with various configurations of electrons creating the next evolutionary Level of matter – a variety of **Atoms**.

Once the atomic Level stabilized, the atoms went through their own evolutionary stages, *combining* to form hydrogen gas molecules that in turn *combined* to form stars, and later stars and planets. On at least one planet (*earth*), molecules of atoms *combined* to form large macromolecules, creating proteins, carbohydrates, nucleic acids, and lipids, that eventually *combined* to form the next evolutionary Level of matter – the **Cell**.

Cells in turn went through their own evolutionary stages developing sensory apparatus, locomotion, a circulatory system, a nervous system, and the ability to *combine*. These advancements lead to the creation of the next evolutionary Level of matter – the **Organism**.

One of the definitions of evolution is the gradual development of something, especially from a simple to a more complex form. Based on observations of these four evolutionary Levels, we might conclude that evolution progresses in stages to *Levels*, and that each Level is the result of the previous Level growing in complexity and *combining*. We could call this process the Combination and Growth Process. But since it has existed for billions of years and seems very precise in both its persistence and direction, if we can find evidence that matter has *evolved* in this way since the very birth of the universe, we could more precisely call it the *Combination and Growth Imperative*, or the CAGI for short. We would only need to show that matter has *evolved* by this process, which is one of the basic tenets of this treatise.

The constant building of complexity has long been observed by others. Physicist Paul Davies tells us, 'The fact that nature has *creative power,* and is able to produce a progressively richer variety of complex forms and structures, challenges the very foundation of contemporary science.'[1.] His book, *The Cosmic Blueprint*, was published in 1989, but the notion is still relevant today.

There is plenty of evidence for the building of complexity, but there is also a force tearing things down. Some believe the force tearing things down is the dominant force and will be the one most influential in determining the destiny of the universe. Why would the universe build into its fabric a force to tear things down?

6.3 The Purpose of Entropy

Thermodynamics is the study of heat and energy. The reader is probably familiar with the first law of thermodynamics: energy cannot be created or destroyed. The second law of thermodynamics encompasses the idea that things and systems are usually wearing down, becoming less energetic and more disordered and random. The term to describe this phenomenon is *entropy*. There are many interpretations of entropy and what follows is only one of many perspectives on the subject.

Disorder implies a prior state of order; you can't have the latter without the former. Throughout the universe, both the building of order and disorder are taking place simultaneously, and for good reason. The Combination and Growth Imperative (*CAGI*) progresses in evolutionary stages to produce Levels of matter that are put together from the elements in the Level that preceded it. In this way, matter is constructed from building blocks that are constantly changing.

A growth in complexity requires structures to be configured in new ways. Building blocks must be available and capable of combining to test which combinations are best suited for the environment in which they reside. The best building blocks are of course those that have already endured the tests of a punishing, hazardous, or frenzied environment, and are then recycled; fragments tossed into physical boneyards or suspended in the Penergy medium to be reused when needed.

Entropy is the universe's way of breaking down complex units into smaller building blocks. Even the most highly complex levels are subject to entropy. If nothing broke down at each level of complexity, we would soon run out of choice building blocks of matter and energy, and further complexity would be stunted. The environment is constantly but subtly changing, requiring constant entropy to produce building blocks with the latest core, survival attributes allowing for adaptation. For complexity to grow effectively, death and decay must be a part of the process. No level or its building blocks is permanent. The building blocks are used by the CAGI process temporarily to further its progress and then recycled. It's just how evolution and the universe work.

6.4 Emergent Qualities

To examine whether the CAGI theory has merit we must determine whether it is feasible for the early elementary particles to have initiated the CAGI by combining in complexity. But Tor particles are simple, spinning tornados of Penergy; how can they possibly grow into the complexity of matter we see today?

Looking at a lone Tor particle, one could not visualize how a small, spinning tornado of Penergy could ever grow into something more complex. We will likely ask the same 'how could this possibly grow?' question at the start of each evolutionary Level because the dominant structures at the start of each Level seem so simple, they show no evidence or promise of combining potential. What we later observe, however, is that each Level evolves new and complex *emergent qualities* that assist it in changing, growing, and combining.

Emergent qualities come about according to Physicist Stephon Alexander when elementary constituents interact to create novel properties that are not possessed by the constituents themselves.[2.] Looking at two hydrogen atoms and one oxygen atom, one could not anticipate their combining to produce a substance such as water. Nor could one anticipate the creation of all the other gases, liquids, and metals that can be produced from other combinations of simple atoms. Emergent qualities are at the heart of the evolutionary process.

The emergent quality in Tors was, of course, its effect on the Penergy medium that created *fields* that gave Tors the capacity to combine into chains, and all particles the capacity to interact. Emergent qualities are important tools used by the CAGI in the advancement of cosmic evolution. Additional emergent qualities will be identified as we identify future evolutionary Levels.

6.5 Cosmic Homogeneity

The next cosmic issue to be examined has plagued cosmology for a long time: Why does the universe today appear so homogenous? Homogenous means something having its components (*like heat, stars, and galaxies*) uniformly distributed throughout.

According to Astronomer Carolyn Devereux, that question is known as the cosmic *horizon problem* (aka *homogeneity problem*) and is one of the most important problems in cosmology.[3.] The leading theory to explain the universe's homogeneity has not been made a part of the Standard Model of Cosmology but is still widely popular. The theory is rather incredible and deserves our examination. The homogeneity issue also addresses the sequence in which the universe evolved, so it is important for our puzzle assembly project to get it correct. Let's examine homogeneity as expressed in the leading theory and how that puzzle segment would appear in the Tor Model.

To overcome the homogeneity problem and other conceptual problems within the BBT, Physicist Alan Guth in the 1980's put forth a radical explanation called *inflation*. He theorized that within two hundredths of the first second of the big bang the universe blew up to a comparatively enormous size very rapidly. Stephen Hawking described it as if a one-centimeter-wide coin suddenly blew up to ten million times the width of our Milky Way galaxy.[4.] Inflation got rid of three conceptual problems all at once: the *Flatness Problem*, the *Missing Monopole Problem*, and the *Homogeneity Problem*. Let's see how the Tor Model takes care of those three problems in a reasonable and less explosive way.

The Flatness Problem: At the big bang the universe would have been under a heavy gravitational influence that should have resulted in a sharp curvature of spacetime, but that is not what we see. Inflation is said to have ironed the sharp curvature out to the flatness observed. In the Tor Model, the universe was not under the influence of gravity at or immediately after its birth, so there was never a sharp curvature to spacetime that needed flattening out.

The Missing Magnetic Monopole Problem: The BBT mathematics predicted the existence of magnetic monopoles, which inflation should have gotten rid of. As shown in 5.2h, the magnetic field arises from the coupling of like-spinning Tors aligned nose to tail, thus necessitating a dipole by design. In the Tor Model magnetic monopoles cannot possibly exist so are not an issue.

The Horizon-Homogeneity Problem: Cosmologists presume the universe is uniform and appears the same from any vantage point within the universe. Over a distance of 300 million light years, from any point in the universe the galaxies and distances between them look much the same, the stars and their contents look much the same, etc. This presumption is so strong it has been named the *Cosmological Principle.* This uniformity, however, is not consistent with the universe envisioned by the Cosmological Model, which does not predict a uniform universe, making the horizon problem a fundamental issue. According to the Standard Model, the star clad galaxies did not start to form until millions of years after the big bang.[5] By this time the universe was too big for it to balance out and have its contents uniformly distributed to the degree we see today.

Imagine a large, cold room with a thermostatically controlled wall heater. The temperature drops, the heater goes on and begins to heat the room, creating hot and cold areas as the heat disperses. Given enough time, the temperature in the room would equalize and be uniform throughout. But what if the room was constantly expanding at a rate faster than the equalizing effect could take place. The temperature could never reach uniformity. This is the scenario scientists have calculated for the early universe. Under the Standard Model's scenario, no matter how far one goes back in time, the universe was never compact enough for a long enough time to achieve homogeneity.[6]

We don't know what started the inflationary expansion or what stopped it. The expansion may solve the homogeneity problem but stretches believability. The theory is complicated, involves undetectable fields, the creation of new, undetected particles (*inflatons*), and has other ad hoc characteristics.[7] This theory may work mathematically but is difficult to believe. Physicist Tony Rothman tells us that the inflaton fields were introduced solely for the purpose of producing rapid expansion and have no observational or theoretical justification.[8] Physicist Sean Carroll reminds us that there is no evidence that inflation ever happened.[9]

Let's explore an alternative scenario for the universe's homogeneity by looking again at the Cascade Effect and SMBH and particle formation. The process begins with the initial speck of pure energy expanding and losing density and breaking up into large swirls with marginal gravitational influence. As detailed in Section 3.3, the

Cascade Effect takes place resulting in smaller and smaller swirls, ultimately creating SMBHs, particles, and the IPM. The newly minted photons and other early elementary particles throughout the universe were free to roam and begin combining. Once the bigger swirls gained sufficient angular momentum and a serious gravitational influence, the free particles were drawn closer. As the swirls developed into SMBHs, their spin caused the surrounding particles and proto stars to swirl, flatten to a disc, and form the structure of our galaxies.

All of this suggests that SMBHs, particles, and early stars were formed at roughly the same time. Most galaxy mass is in the surrounding particles and stars, and the SMBH generally represents only 1% of the galaxy's total mass, suggesting the SMBHs likely drew very little particle mass inside. Eventually, the SMBHs would have attracted exactly the right amount of gas and star material to support their gravitational field.

This scenario seems consistent with the findings of scientists looking at some of the oldest galaxies that date back twelve billion years or more. As reported in an article dated 1/27/2021 in *Scientific American* entitled *Giant Galaxies from Universe's Childhood Challenge Cosmic Origin Stories,* some of those very old galaxies are too large to have formed in the customary way of being built up from hydrogen gas and star debris. This article is consistent with the one discussed in Section 3.3. Both articles report that scientists are challenged to find a reasonable explanation for how those galaxies could have become so big, so soon after the big bang.

To explain both the size and age of the galaxies only requires the proper timing for particles and particle combinations to have come into existence just as the pure energy blackholes were gaining a gravitational influence. By this time the universe was uniform in its contents and remained homogeneous right up until the time the

Standard Model sees atoms created and photons decoupling from matter creating the Cosmic Microwave Background.

The cascade effect accounts for the distribution of the galaxies but does not explain the ultra-consistency of the temperature of the CMB. Whatever process brought the energy level of those photons down to a very consistent temperature may have had little to do with the homogeneity of the rest of the universe. We will examine the issue of photon temperature and the CMB in Chapter Seven.

The stage is set for Tryk particles to combine into larger more complex particles that will constitute our Sub-As, which will be examined next.

Resolution of Cosmological Issues – Ch.6

☒ Explanation of Work Energy
☒ Foundation of Quantum Mechanics
☒ Magnetic Monopole Mystery
☒ Missing Anti-Matter Mystery
☐ Neutrino Oscillation
☒ Origin of Cosmic Web
☒ Origin of Dark Energy
☐ Origin of Dark Matter
☒ Origin of EM field
☐ Origin of EM Waves
☒ Origin of Fields
☒ Origin of First Stars
☒ Origin of Force
☐ Origin of Gravity
☐ Origin of Mass
☒ Particle Decay
☐ Particle Evolution Sequence
☒ Particle Tunneling
☒ Purpose of Entropy
☒ SMBH Creation
☐ Space & Time Relativity
☐ The Fine-Tuning Problem
☒ The Flatness Problem
☐ The Hierarchy Problem
☒ The Horizon Problem
☒ The Magnetic Field
☐ The Measurement Problem
☐ The Reality Problem
☒ Virtual Particles
☒ Wave-Particle Duality Resolution
☒ What Drives Cosmic Evolution
☒ What is Energy
☐ Why Gravity so Weak
☐ Why Photons are Massless
☒ Why Three Generations

The absence of a true measure of spin and angular momentum for electrons and quarks is better understood if they are recognized as composites rather than elementals.

Chapter Seven
7.0 Configurations: Photon & Neutrino

Tryks are the stable, second level of matter. To reach the next stable level, Tryks will combine into the complex forms we know of as subatomic particles (Sub-A's). Particles combine in two ways. Firstly, due to a natural attraction inherent in their fields such as like-spinning Tors combining nose to tail to create a chain, or the natural attraction between electrons and protons to create atoms. Secondly, two or more particles can also combine by sharing parts of themselves, such as atoms sharing one or more electrons to create a molecule. In our examples, Tryks combine using field energy that allows for natural attraction.

Either of these methods will hold two or more particles together but other scenarios may be possible. According to the Standard Model, protons are made up of three quarks that are held together by a strong force that involves both sharing and natural attraction. Such a force may represent an entirely different way of combining, but its existence is only theoretical. In the Tor Model, all forms of combining are ultimately due to the attractive and repulsive interactions between particle fields.

The strength of the field force holding particles together is called *binding energy*. It takes at least the amount of the binding energy to pull apart the constituent elements of a composite particle. Tors spin on their own axes and create a very strong field, making the binding energy of Tor and Tryk Chains quite high. We may never build a particle accelerator with sufficient energy to overcome the binding energy to break matter down to even the Tryk level.

7.1 Arguments for Composite Particles

Angular momentum is the rotational analog of linear momentum, an attribute of a body measured as the product of mass and velocity. Particles spin and therefore possess some measure of *angular momentum*. In Quantum Theory (QT), fundamental particles do not spin on their own axes. Their spin cannot be measured directly but is measured indirectly using *projections*. QT particles are therefore said to have an *intrinsic* amount of angular momentum. According to Physicist Wouter Schmitz, '...we really do not know what [*quantum*] spin is exactly.'[1.]

In the Tor Model, the only particles spinning on their own axes are the Tors, Tryks, and Photons. Electrons and quarks do not spin on their own axes because they are composites with complex configurations. As composites, comprised of smaller particles in various spin orientations, electrons and quarks spin on a center-of-mass axis. That is probably why as composites, they possess only a poorly defined 'intrinsic' spin. The absence of a true measure of spin and angular momentum for electrons and quarks is better understood if they are recognized as composites rather than elementals.

In Quantum Theory there is a mystery as to why it takes a fundamental particle two complete revolutions to return to its original position. Composite particles may be the answer. Scientists have noted that spinning gyros in multiple orientations are stable if their moments of inertia are the same. If different, however, one of the orientations becomes unstable allowing the system to tumble in the unstable direction. Given the multiple spin orientations within a composite particle comprised of many Tor-Chains, instability in one of its orientations is inevitable, causing the particle to not only spin but tumble. If it tumbled one-half turn for every complete spin rotation, it would then take two rotations for it to complete a full tumble and return to its original position.

Quantum Theory ascribes this later mystery to the existence of an unusual *spin-space*. Physicist Bruce Schumm asks, "So the question is, what exactly is spin and this oddly construed spin-space in which it lives?" He subsequently answers, "We don't really have a clue about the physical origin of spin."[2.] Again, these mysteries are better understood if quarks and electrons are recognized as composites rather than elementals.

7.1a Support for Composite Particles

As demonstrated, spin is a prevalent and important characteristic throughout the universe. Penergy spinning itself into simple elementary particles seems to be the most likely way for particles to have come into existence. Simple elementary particles that then combine and evolve into Sub-A's seems quite feasible, especially given the observations and deductions made so far. Many physicists tell us our Sub-A's could feasibly have a substructure. Physicist Harald Fritzsch tells us that though there is no experimental evidence for it yet, it is possible that the electron possesses an inner structure,[3.] and that quarks and leptons might consist of smaller building blocks.[4.]

Physicists John Barrow and Joseph Silk tell us that physicists would not be surprised to discover ultimately that quarks and leptons have internal constituents.[5.] Physicist David Lindley relates that, "Quarks and Gluons may not be truly elementary but built from even more arcane ingredients."[6.] Physicist Don Lincoln says that, "... it is at least plausible to consider the possibility that quarks and leptons are composed of even smaller particles held within them."[7.]

Physicist Jon Butterworth tells us both categories of matter particles, quarks and leptons, may contain smaller constituents.[8.] Chemical Physicist Michael Munowitz relates that quarks may not be the indivisible, end-of-the-line building blocks, but they are probably only a step or two away.[9.]

Stephen Hawking told us that at very high energies we might expect to find several new layers of structure more basic than the quarks and electrons that we now regard as elementary particles.[10.] Dr. Jorge Cham and Physicist Daniel Whiteson convey that there is no proof the particles we see today, electrons and quarks, are the most basic building blocks in the universe, but are [*merely*] the smallest bits of matter we have seen so far.[11.] According to Astronomer Chris Impey, the Standard Model leaves open the possibility that there's a deeper level of structure.[12.]

Obviously, though there is yet no evidence for it, many physicists feel the existence of constituent particles comprising our Sub-A's is at least plausible. The following points support that notion.

- As we have seen, composite particles can account for the missing anti-matter. The anti-matter is not only a part of composite particles, but an essential part. As we progress in building a picture of our universe, the integrated anti-matter particles will again prove to be an important element in that construction.

- Composite particles could be the underlying source of *magnetism* and *charge*. As discussed in Sections 5.2b-c, Tor-Chains with all their spin axes aligned create a magnetic field and the attraction of opposite spinning Tors creates an electric field. Consequently, it is the combination of Tors that creates both magnetic and electric elements. Our examination of quark configurations soon will demonstrate how the combination of elements in composite particles creates various amounts of charge.

- Composite particles imply the existence of a core particle spinning on its own axis such as the Tor. The validity for the existence of a particle like the Tor is implied in many of the hypotheses discussed so far. The spinning of the Tor particle may also be the source of mass, as examined in Appendix #1. This is in addition to the Tor being the source of the magnetic, electric, and gravitational fields (*discussed in Appendix #3*). It seems like quite a bit to expect from a single particle, but it makes more sense for these attributes to have *evolved* from the characteristics of the most elementary particle than for those fields and forces to have simply popped out of a hypothetical inflaton field.

- In the BBT, there is no evolutionary explanation for the origin of fundamental particles; no explanation as to how the high-density speck of energy suddenly changed to multiple configurations of light density particles. The creation of a core particle that combines to produce composite particles infuses natural selection into the evolutionary process, which seems like a much more credible influence on particle composition than 'it just happened'.

- It is observed that particles cannot spin on their own axis because the amount of measured angular momentum they produce would cause them to exceed the maximum spin rate. That high level of measured angular momentum may be due to it being the sum of angular momentum of all the composite elements making up the particle.

- Scientists tell us that all charges are rational multiples of a basic unit of charge. The fact that there are three Sub-A's (*electron, up quark, and down quark*) with the same type of electric charge but with different values makes it probable they share a common, underlying structure. Jorge Cham and Daniel Whiteson tell us that because the charge of the electron perfectly matches the charge of the proton but is opposite, it is another sign there are deeper components underlying those particles.[13.] The ease in which various charges can be configured is a compelling argument for composites.

- As mentioned in the previous section, the mysteries of particle spin and angular momentum are more easily understood if electrons and quarks are recognized as composites rather than as elementals.

All these facets weigh heavily toward today's Sub-A's being composites of more elementary particles. In the Tor Model, those elementary particles are Tors and Tryks, but there may be deeper levels and they could look much different. I am only arguing for their presence, not their appearance.

7.2 Particle Configurations

In the Tor Model, Sub-A's are comprised of Tryks, which represent a very important evolutionary Level. Tryks brought important emergent qualities to the universe. The fields surrounding the Tryks reflect the fact they are comprised of both Tor-Chains creating a magnetic field and a combination of matter and antimatter creating an electric field. This combination of emergent qualities created the electromagnetic field, which gave those particles the very important ability to absorb and emit photons. Photon absorption gave particles built from Tryks the capacity for different energy levels to promote various interactions and combinations. These emergent qualities not only assisted the Tryks in combining, but they will prove essential for the growth of complexity for many Levels of matter well into the future.

The cosmic environment at this point was likely a hot frenzy of activity, perhaps too busy for complex matter to form. But as the universe expanded and cooled, the Tryks would have begun combining. Those new combinations would have to be resilient to survive. Many trial combinations of Sub-A's would have been created and destroyed, testing the combining strength of the Tryk. This slow, testing, transitional phase occurs between all Levels; it is simply a part of the CAGI's evolutionary process.

To explore particle characteristics, Sub-A's are smashed together at high-speed using huge particle colliders such as the twenty-seven-kilometer, circular, Large Hadron Collider (LHC) located underground near Geneva, Switzerland. The collision creates a knot of Penergy big enough to bring into momentary existence heavy, unstable particles that immediately decay into lighter particles. Scientists have developed detection equipment to record the behavior of these lighter particles, which gives the scientists a basis for theorizing the particle's existence, characteristics, and relationship to other particles.

Particles have been categorized. Due to shared features, electrons and neutrinos are called leptons. Quarks are in their own category, and there is also a category called Bosons, which includes the photon and all virtual force-carrying particles. Photons are considered energy carriers that provide matter particles with the requisite energy to move, interact, and combine.

Some massive particle creations are so fleeting they have not yet been seen but are theorized based on the behavior and characteristics of collision debris. Scientists have identified only four, naturally existing stable particles and their anti-particles: the quark, neutrino, electron, and photon. They have also either seen or theorized a host of many unstable particles and anti-particles. I will examine the configuration of only the stable, third generation of particles. Again, it is not my intent to argue the specific configurations for particles, only the possibility for them.

All particles are made from Tryks, and their composition is likely close in configuration, which explains why each can be created from the other. Possible Tryk combinations are near infinite. The scheme I chose uses the Tryks illustrated in Figure 5.4.

7.2a EM Waves

I will next address the possible configuration of the photon, but first need to begin with a seemingly crazy idea: photons may not be EM waves, per se. Photons give off the same wave signature as EM waves and both travel at light speed, so they appear to be one and the same, but perhaps not.

True EM waves originate from devices such as radio transmitters. The waves are produced by a specialized transducer that races electric current up and down an antenna. Free electrons in the antennae act as a media propagating the waves. The movement of charged particles creates a field that radiates out from the antenna at the speed of light, which are true EM waves. Those waves spread out in all directions, like a wave created by dropping a rock into a pond. The wave's energy (*frequency*) is controlled by the transmitter, radiating out the sides of the antennae at the same frequency as the variable voltage applied to it. The EM wave propagates with a certain electric and magnetic signature. Those waves are not single photons, as evidenced by the fact they spread out evenly in all directions.

The photon has the exact same field/wave signature. The difference is that photons do not spread out as they propagate.[14.] Photon trajectories create EM signature waves in the IPM, but they only propagate out from the photon's path. Photons create EM-like waves because of the photon's configuration, discussed next.

7.3 Photon Configuration

While likely all our primary particles evolved roughly at the same time, the photon likely came late to the party. The photon is not utilized to build complexity physically but is nonetheless very instrumental in its construction. Once Tryks began to combine into heavier particles it must have been difficult to move, attach, and detach particle parts without some form of energy to assist in those processes. Consequently, once a photon accidentally came into existence and proved itself essential to those processes it would have quickly become a dominate particle.

Photons are customarily referred to as particles of light. They are absorbed and emitted by matter at specific wavelengths, which then enter our eyes creating impulses that are interpreted by our brains giving us vision. They exist in a span of energy, ranging from low energy radio waves to very high energy gamma waves. Except when passing through a clear medium, such as air, glass, or water, they travel in a straight line and at light speed, roughly 3.0×10^8 meters/second.

Photons are never at rest, so there is no way to measure their rest mass, but they are believed to be massless. They can be absorbed and emitted by charged particles but otherwise exist only in constant motion within the Penergy medium. They are considered both a particle (*photon*) and a wave but may only be simply a particle with wavelike characteristics, as argued in Chapter One. When absorbed, their energy is transferred to the particle, boosting its kinetic energy. Due to its configuration, a photon is its own anti-particle.

7.3a Photons According to the Tor Model

A photon spins on an axis parallel to its direction of travel as evidenced by the torque effect it has upon impact with a target. Like other particles, its spin creates a personal field with wave-like characteristics, including wave frequency. The energy of a photon is expressed in terms of the equation E=hf. In this equation, the h is Planck's Constant [6.626 x 10^{-34} Joule Seconds], a term that appears in nearly all quantum equations, and f is frequency. A Joule Second expressed in international standard measurements is kg m^2/s, which is an expression of angular momentum - the measurement for spin-momentum. So, photon energy (E), as expressed in terms of hf, means angular momentum times a frequency. If photons are proven to be particles and not waves, that frequency would be interpreted in terms of spin rate that results in a commensurate wave frequency. That means that a photon's energy is directly related to the rate at which it is spinning, which in turn is creating its personal field with the same frequency, making a photon's energy directly proportional to its spin rate and wave frequency. The photon's spin rate is established when it is created but can be changed by outside influences.

Photons travel at the maximum speed limit of $3.0x10^{-8}$ m/s. Their travel speed is due solely to their relationship with the Penergy medium, which is why they always measure at the same speed irrespective of the energy level of the photon or motion of their source. Because photons travel at the same speed as massless EM waves, photons are considered massless. They are known to possess momentum and be capable of instilling a collision impact on other particles, strongly suggesting they possess a mass-like content. (*How a real particle can be 'massless' is examined in Appendix #1.*)

7.3b Speculation on Source of Photon Energy

A photon spins around a central axis as illustrated in Figure 7.1. The rate of that spin is a function of the fields of the adjacent Tryk-Chains that cause the chains to spin around each other. The strength of those fields is in turn a function of the distance between the chains – the closer the chain separation distance, the faster the spin rate. The distance between the Tryk-Chains is established by the density of the Penergy immediately surrounding the photon when it was created. The heavier the density, the more it compresses the local area, bringing the Tryk-Chains together. The configuration quickly reaches an equilibrium and goes on its way. The degree of Penergy density involved becomes a part of the photon's overall field, and once established, the particle's equilibrium maintains the photons spin rate and configuration. Depending on the Penergy density conditions when the photon is created, it could have a wide range of spin rates, accounting for its wide range of energy levels. When a photon decays, its mass and spin energy are turned back into a knot of Penergy. At high energy levels, the photon could decay and be reconfigured into any particle pair commensurate with that energy level.

To satisfy all the known traits of a photon, the configuration need not be extraordinary. It could be quite simple, similar to a neutral, two-Tryk combination, as illustrated in Figure 7.1. The challenge, of course, is to create a composite that is neutral in charge without the like-components annihilating. The configuration shown in 7.1a is two Tryks that came together with different lengths but with their compatible fields drawn to each other. They unite at their Tor-Chains rotating around each other creating a strong, high-spin bond. The uneven chain lengths creates an instability that causes excess Tors to be culled from each chain consistent with it creating its own level of equilibrium. The spin of this configuration around a central axis would result in a complex field with an EM signature that could create a mild attraction between the photons. The only difference between photons is their energy level, so two photons with the exact same energies could annihilate but they don't because the particle is neutral in charge and lacks a strong attraction to other photons.

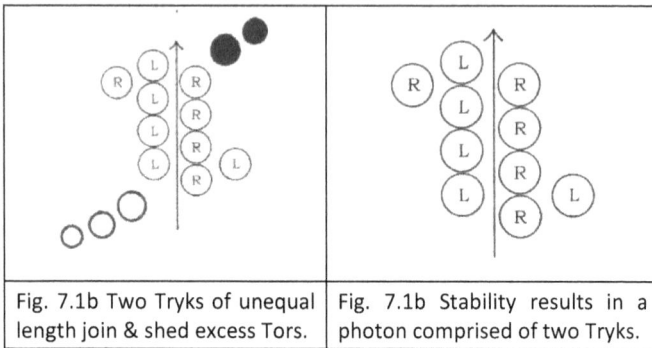

Fig. 7.1b Two Tryks of unequal length join & shed excess Tors.	Fig. 7.1b Stability results in a photon comprised of two Tryks.

Figure 7.1 Possible photon construction and configuration.

A photon is thought to be an electromagnetic wave because of its configuration. Propagating in the direction of its Tryk-Chains, it displays a magnetic field, while the opposite-spinning Tors encircling the chains present an electric field that is perpendicular. Propagating in this fashion at light speed, the personal field of the photon would look like an EM wave. It is the same EM wave created by antennas because the electrons racing up and down the antennae are chains of Tryks giving off the same wave signature.

Photons are said to be absorbed and emitted from charged particles. Once proven to be a particle and not a wave, the idea of *absorption* is problematic. To be absorbed would mean it would have to decay from a particle state to an amorphous knot of Penergy that then somehow becomes a part of the Tors or Tryks making up the charged particle. I don't see how particles already spun into existence could *absorb* additional Penergy, either massive or amorphous. If they could, they would constantly absorb all kinds of available knots of Penergy. Consequently, I doubt the photon is absorbed per se, so it must continue to exist as a particle. Perhaps, the 'absorption' is the photon attaching itself to the end of the charged particle's Tor-Chain due to the photon possessing a compatible field with the charged particle.

The description of the photon's movement also serves to describe the relationship between magnetism and electricity. Scientists have long known that electricity moving through a wire induces a magnetic field. This makes sense if one imagines the spinning electrons moving through the wire headfirst, parallel to their axes of spin. All of those moving, like-spinning electrons in a head-to-tail arrangement would mimic a Tor-Chain and therefore induce a magnetic field.

Likewise, consider a generator. When a magnet comprised of many particles with aligned spin-axes is rotated, such as an armature, the rotation creates an unequal balance of charged particles, inducing an electric field, causing the electrons to move through the adjacent wires. Each of the fields in turn induces the other.

7.4 Connector/Neutrino Configurations

Combining Tor-Chains with Tor singles may have given rise to a problem of how to make those connections stable, requiring a special *connector* to have evolved. As previously observed, the strong coupling capacity of the Tor field encouraged a nose to tail coupling creating Tor-Chains. The addition of an opposite-spinning Tor encircling the chain could cause a lengthy Tor-Chain connection to wobble, weakening the field connection strength between the connecting Chains. As the Chain links grew longer and the wobbles grew more erratic, the magnetic field strength of the Tors holding the Chain connections together would have been challenged. The wobble may not have been a problem for short Chains but a serious problem for long Chains.

This limitation to Tryk-Chain length could have been overcome if there had evolved another independent particle that could serve as a flexible connector (○) between the Tryks. In this way the Tryk-Chain connection would not be completely dependent on the magnetic force between Tors, allowing for stable Tryk-Chains of any length. A connector could also be configured to connect a positive Tor-chain to a negative Tor-chain.

Such connectors would likely be very small and neutral in charge. They might exist without our even knowing it. Being small, neutral, and likely near massless, in the collision debris calculations in colliders and accelerators they could easily be mistaken for neutrinos, or they may even *be* neutrinos. Let's examine the prospect of the neutrino being a connector.

7.4a Neutrinos as Connectors

Neutrinos are one of the most abundant particles in the universe. They have neutral electric charge, are nearly massless, and seldom interact with other matter.

In the Tor Model, neutrinos may serve as connector particles for electrons and positrons. Neutrinos only seem to appear in conjunction with the emission of an electron or positron due to particle decay or during the fusion process at the heart of a star. Every time an electron or positron is spit out, so is a neutrino. This close tie suggests a possible physical connection between the particles.

The neutrino may be a *Majorana* – a particle configured like its anti-particle. If the neutrino is its own anti-particle, it would have either a unique configuration or unique qualities. As such, it may serve as a universal connector allowing for the connection between all sorts of Tryk-Chains and other composite particles.

A particle's mass may be linked to the particle's constituent configuration, as discussed in Appendix #1. For the neutrino illustrated in Figure 7.2, nearly all the constituent Tors have their axes oriented in the direction of travel, but the central mass misses that parallel orientation just enough to give the neutrino a small inertial mass.

This configuration may explain how the neutrino can oscillate - change mass/identity while in flight. The oscillation results when the neutrino suffers changes in its structure, which means the Tau, Muon, and Electron Neutrinos are all the same particle with changeable configurations. For instance, the neutrino pictured in Figure 7.2 could, while in flight, pick up an electron at one end and a positron at the other, changing its configuration and mass content without changing its neutral charge or other characteristics.

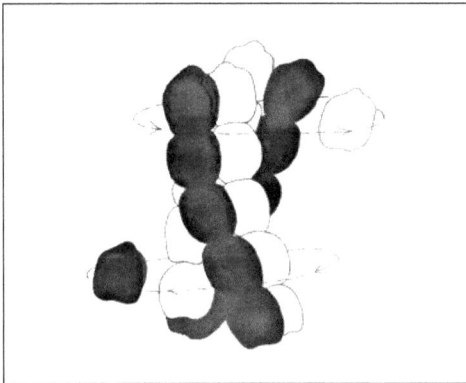

Fig. 7.2 This flexible configuration would allow the Neutrino to serve as a connector between left or right spinning Tor/Tryk-Chains.

The neutrino could be an even number of Tor-Chains in a bundle. The four Tor-chains with different lengths comprising a bundle would be drawn to and rotate around each other due to electric field attraction. If the bundle was joined by a single Tor encircling the end of the bundle, it would create a charge imbalance that would allow it to connect to either a positive or negative Tryk-Chain. The connection could be accomplished by either the respective fields joining, or the single Tor being shared between the connector and the Tryk-Chain, like electrons are shared by atoms to create molecules.

The bundle's attraction to a single Tor would give the neutrino a momentary electric charge differential allowing for the absorption of photons, changing its energy and momentum, and giving it the attributes of a more massive particle. Having both left and right spinning Tor-chains exposed at its ends means the neutrino could attach to either kind of Tryk-chain, making it an ideal, flexible, universal connector.

Neutrinos are known to only spin to the left, and anti-neutrinos to the right. This suggests they may be a Majorana since they appear as the same particle whether viewed from nose or tail. The neutrino in Figure 7.2 is moving upward, showing a left spin. Technically, a right spinning neutrino is its anti-particle, but again depending on how its configured it could spin oppositely without a change in charge value, making it appear a Majorana.

The story of Sub-A configurations for the Quark and Electron will continue in Chapter Eight.

Resolution of
Cosmological Issues – Ch.7

☒ Explanation of Work Energy
☒ Foundation of Quantum Mechanics
☒ Magnetic Monopole Mystery
☒ Missing Anti-Matter Mystery
☒ Neutrino Oscillation
☒ Origin of Cosmic Web
☒ Origin of Dark Energy
☐ Origin of Dark Matter
☒ Origin of EM field
☒ Origin of EM Waves
☒ Origin of Fields
☒ Origin of First Stars
☒ Origin of Force
☐ Origin of Gravity
☐ Origin of Mass
☒ Particle Decay
☐ Particle Evolution Sequence
☒ Particle Tunneling
☒ Purpose of Entropy
☒ SMBH Creation
☐ Space & Time Relativity
☐ The Fine-Tuning Problem
☒ The Flatness Problem
☐ The Hierarchy Problem
☒ The Horizon Problem
☒ The Magnetic Field
☐ The Measurement Problem
☐ The Reality Problem
☒ Virtual Particles
☒ Wave-Particle Duality Resolution
☒ What Drives Cosmic Evolution
☒ What is Energy
☐ Why Gravity so Weak
☐ Why Photons are Massless
☒ Why Three Generations

Given that neutrons naturally decay into protons in a matter of minutes, it makes sense for neutrons to have evolved after the hydrogen atom evolved, which allowed neutrons to readily join a nucleus before decaying.

Chapter Eight
8.0 Configurations: Quark, Electron, Nucleus & the Atom

According to the Standard Model, protons and neutrons are composite particles comprised of three quarks and a sea of virtual gluons that by exchange hold the quarks together. Quarks and gluons have not been isolated or observed directly but are theorized to exist based on experimental evidence.[1] Quarks are believed to be elemental, are never found as a single unit, and are normally found paired together in twos or threes. They are believed to carry electric charges but only in $1/3^{rd}$ or $2/3^{rds}$ of the whole $3/3^{rds}$ charge value of the electron.

8.1 Constructing Quark Charge Values

Before looking at possible quark configurations, let's examine how quarks possibly obtain their electric charge values, which may influence their configuration. In the Tor Model, quarks are not elemental but are comprised of Tryks, and their fractional charges can be easily constructed from those Tryks. The strength of the electric force is measured in charge value, but no particle possesses the inherent attribute of charge. Charge is simply a measure of the strength in the *interaction between the fields* of Tors. The greater the differential in the number of left and right spinning Tors, the greater the strength of the electric charge. To measure the differential, we can place a theoretical value on each Tor.

As observed, the measure of quark charge is $1/3^{rd}$ and $2/3^{rds}$ of the electron. For the sake of demonstrating how these partial electric charges are possible at the Tryk level let's give each individual Tor particle a hypothetical $1/9^{th}$ charge value. Accordingly, right-spinning (*negatively charged*) Tors would have a $-1/9^{th}$ charge value, and a left-spinning (*positively charged*) Tor would have a $+1/9^{th}$ charge value. Again, these values are ascribed for demonstration purposes only.

With these values and terms, we can construct all sorts of particles with all sorts of electric charge values both positive and negative. Because Tryks contain both positive and negative Tors, each pair of which would add to zero charge, we only need to calculate the charge value based on the net number of charges.

To get the net number of charges for each Tryk we add up all the charges. As shown in Figure 8.1a, a Positive Tryk would have a gross charge value of $+4/9^{ths}$ for the four positive (*left spinning*) Tors, and $-1/9^{th}$ for the one negative (*right spinning*) Tor, giving it a net charge of $+3/9^{ths}$, or a $+1/3^{rd}$ overall charge value.

As shown in Figure 8.1b, a Negative Tryk would have a gross charge value of $-4/9^{ths}$ and $+1/9^{th}$, giving it a net charge of $-3/9^{ths}$, or a $-1/3^{rd}$ overall charge value. In summary, the electric charge value of any particle is determined by the net differential in the positive and negative Tor count making up the particle.

Fig. 8.1a. Charge values of a Positive Tryk: net +3/9ths or a +1/3rd overall charge.

Fig. 8.1b. Charge values of a Negative Tryk: net −3/9ths or a −1/3rd overall charge.

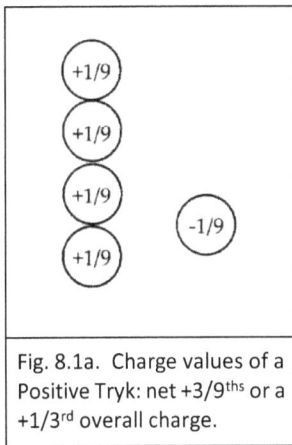

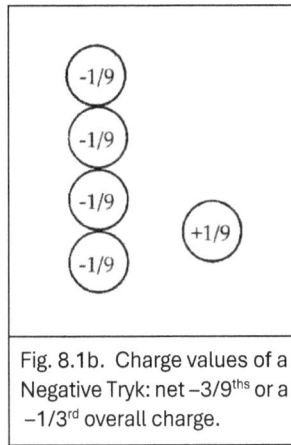

Figure 8.1 Charge values of Tors added together to give a net charge value of a Tryk.

The values given above are arbitrary and are used only for the sake of example. It is not my goal to prescribe a specific charge value for the Tor but only to show how charge values might be distributed between the applicable particles.

8.2 Quark Configurations

Using these Tryk charge values we can now easily create a composite particle with an Up Quark charge value of +2/3rds by joining two positive Tryks into a single chain as illustrated in Figure 8.2a. The anti-particle of this simple illustration would have the mirror image configuration and carry a −2/3rds charge, as illustrated in Figure 8.2b.

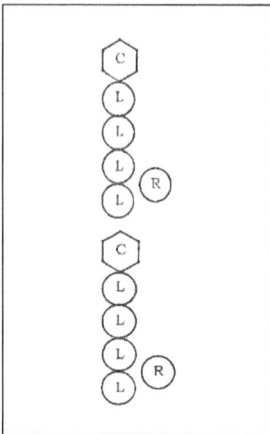

Fig. 8.2a. Two Positive Tryks join to create a +2/3^{rds} charge quark.

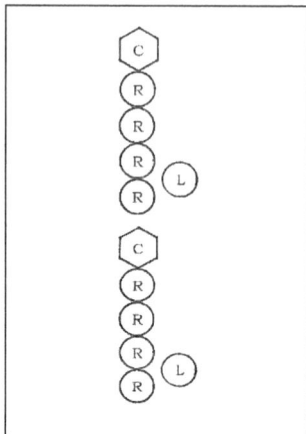

Fig. 8.2b. Two Negative Tryks join to create a -2/3^{rds} charge anti-quark.

Figure 8.2 Representations of a simple up quark configuration and its anti-particle.

8.2a Complex Quark Configuration

A quark configuration might be as simple as Tryks hinged together by a connector, but it is possible that the tenaciousness of the CAGI created much more complex particles. A down quark for example might be made up of a chain of electrons, positrons, neutrinos, universal connectors, and positive and negative Tryks. An example is illustrated in Figure 8.3, which is a triangular shaped particle with corners joined by two-Tryk combinations.

This configuration would have a net -1/3rd charge value, which is the exact charge value of the down quark. Should it spit out the Electron/Neutrino couplet at its corner, it would then have a +2/3^{rds} charge value, becoming an up quark, which is the process called beta decay that changes a neutron to a proton.

Likewise, should the up quark spit out the Positron/Neutrino couplet at its corner, it would then have a -1/3rd charge value, becoming a down quark, changing a proton to a neutron. The ease at which particles decay into other particles in this fashion makes the argument for composite particles compelling. This down quark configuration is not definitive and is only one of many possible configurations for a quark made from Tryks.

With a configuration like that shown in Figure 8.3, the constant exchange of a single electron or positron between the three quarks might create a sufficient charge imbalance to hold them together much the way electron exchange creates a charge imbalance that holds molecules together.

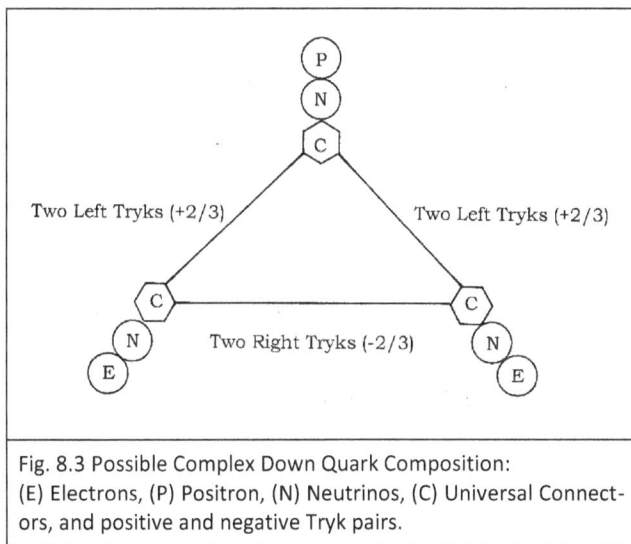

Fig. 8.3 Possible Complex Down Quark Composition:
(E) Electrons, (P) Positron, (N) Neutrinos, (C) Universal Connect-ors, and positive and negative Tryk pairs.

Figure 8.3 This Quark configuration allows for a simple decay to convert to either an Up Quark or Down Quark.

8.2b Simplest Quark Configuration

Let's see how a very minimum quark configuration might work. A decayable $+2/3^{rds}$ Up Quark would only need at a minimum two positive Tryks, an electron, positron, and accompanying neutrinos: a

+2/3, -1, +1 combination. If a decay resulted in the removal of the +1 positron, the result would leave a $-1/3^{rd}$ down quark. Likewise, if we started with a decayable $-1/3^{rd}$ down quark, we would only need at a minimum one negative Tryk, an electron, positron, and neutrinos: a -1/3, -1, +1 combination. If a decay resulted in the removal of the -1 electron, the result would leave a $+2/3^{rds}$ up quark.

If either quark decay was initiated by the absorption of a high energy photon that immediately decayed into an electron-positron pair, we would have all the ingredients necessary for the quark decay. The decay would leave a Tryk-Chain connected to either an electron or positron, while ejecting an electron or positron and its neutrino connector, depending on the type of decay.

Again, I am not suggesting that any one specific configuration is how quarks are comprised or held together. Obviously, there are many options possible and the configurations discussed are only to demonstrate some possibilities. With some head scratching, new experiments, and a dab of mathematics, an ingenious physicist or cosmologist will someday figure it all out.

8.2c Meson Configuration

According to the Standard Model, a meson is a particle comprised of two quarks. The nuclear force binding the protons and neutrons together to form atomic nuclei is theorized to originate from the exchange of a meson called a Pion.[2.] The pion exchange allows protons and neutrons to momentarily exchange identities, helping to overcome the repulsion between the like-charged protons.

A meson is comprised of a quark and anti-quark. In the Standard Model, the mirror-image quarks contain a different hypothetical color charge, so they don't annihilate. A meson can also be comprised of a quark and anti-quark from different generations. Independent mesons are, however, unstable and suffer an immediate decay when created in a shower of collider particle debris.

The Tor Model recognizes the momentary existence of meson-like particles that may be created from various quarks from different generations, but it offers no specific configuration for this unstable, two-quark meson.

8.3 Electron Configuration

According to the Standard Model, the electron is an elementary particle without constituent parts, but there is evidence that could be interpreted otherwise. Physicist Frank Wilczek reports that electrons can be fractured into smaller particles with partial charges through the *fractional quantum Hall effect.* But these are not real particles but only elements that act like independent particles, called quasiparticles. They are created only under precise laboratory conditions.[3.] This research, however, suggests the possibility that electrons could have an inner structure, even if it's only manifested under a very limited and extreme experimental environment.

In the Tor Model, the electron is a composite particle made up of Tryks. An electron carries a whole 9/9[ths] negative-charge value, which means that using the working values we established already it has a differential of nine negative Tors. Three negative Tryks in a chain would fulfill the requirement as illustrated in Figure 8.4a. The positron (*anti-electron*) would of course have the mirror-image configuration as illustrated in Figure 8.4b. The electron configuration is likely much more complex, but for illustration purposes a three Tryk-Chain will be used.

The electron illustrated above is made from a *chain* of Tryks. One might consider creating an electron from three unconnected, independent Tryks. That would seem equally feasible, but it is not possible due to the *Pauli Exclusion Principle*. That principle says that no two like fermions (*electrons or quarks*) can together occupy the same quantum state. We need not go into the principle further but know that if comprised of three Tryks, they are more likely in a chain rather than independent units, unless each unit possesses a unique trait giving each a slightly different identity.

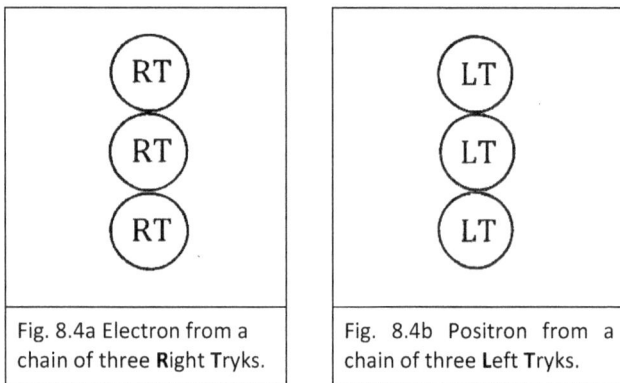

Fig. 8.4a Electron from a chain of three **Right** Tryks.	Fig. 8.4b Positron from a chain of three **Left** Tryks.

Fig. 8.4 Possible configurations of an Electron and Positron.

Even without the existence of a connector, three Tryks may have been able to join nose to tail to create a strong chain. We have concluded that due to the very strong field strength of the Tor, in the early universe the magnetic force could reasonably overcome the electric force if the Tors were aligned nose to tail. Scientists have also observed this to be the case in experiments involving two otherwise repulsive electrons coming together nose to tail to create a combination called a Cooper Pair.[4] This phenomenon came to light while exploring theories of superconductivity (*the flow of electricity without resistance*). Again, it only happens in extreme laboratory conditions. We know, however, that the strength of forces bringing particles together can change under different energy conditions, and the early universe with a declining Penergy density certainly provided the opportunity for different energy conditions.

8.3a Boson Configuration

According to the Standard Model bosons are whole-integer-spin particles and are mostly virtual particles that are theorized to facilitate the absorption, emission, creation, or decay of Sub-A's. The group is comprised of photons, gluons, gravitons, W & Z particles, and the Higgs. For the virtual particles it means they exist only momentarily, flitting in and out of existence. Their hypothetical existence requires them to borrow energy from the vacuum of space, but as shown in Section 5.1, the vacuum apparently does not possess the requisite energy to support their number.

The Tor Model recognizes their possible existence but believes their importance in our universe is uncertain. The Tor Model has an alternative explanation for the source of force and how particles interact and decay, so the possible makeup of these hypothetical particles will not be addressed.

8.4 The Nucleus and Atom

Based on both theory and experimental results, the proton is comprised of two up quarks and one down quark. Those quarks have charges of $+2/3^{rds}$, $+2/3^{rds}$, and $-1/3^{rd}$ respectively, giving the proton a +1 net charge. The reader may recall from high school science class that each element on the periodic table is distinguished by the number of protons in the nucleus, and that two or more protons, because they carry a like charge, will naturally repel each other. Quarks being held together by sharing gluons theoretically spill-over to protons sharing gluons, helping to overcome their natural repulsive force. Apparently, overcoming that repulsive force between protons was insufficient for the gluon and pion exchanges alone, necessitating the involvement of another particle - the neutron. Let's examine how a neutron and the nucleus may have evolved.

8.4a The Nucleus According to the Standard Model

The following points theorize protons and neutrons under the Standard Model. They were created within the first seconds of the big bang and came together to create nuclei during a period/process called Primordial Nucleosynthesis. Unconfined neutrons decay in less than fifteen minutes. In the initial high energy conditions following the big bang, protons and neutrons could exchange identities converting back and forth into each other, but these conditions were short lived, coming to an end as the universe expanded and cooled. Once the conversion stopped, the universe had only fifteen minutes to create the nuclei, otherwise it would have run out of neutrons making it unable to create atoms larger than hydrogen.[5] The protons and neutrons immediately began combining to create nuclei with different configurations, primarily hydrogen, helium, and some of their isotopes. All the existing neutrons were taken up into nuclei within a few minutes of the big bang.[6]

This sequence is part of the Hot Big Bang Theory. That theory may be feasible mathematically but as argued all along, it makes more sense for complex particles to have evolved from something smaller, and for each aspect of evolutionary development to follow from a reasonable cause and effect. Let's continue with the story of how nucleons evolved, but from a more natural, evolutionary sequence.

8.4b The Nucleus According to the Tor Model

A neutron is comprised of two down quarks and one up quark, giving it a neutral (*-1/3, -1/3, +2/3*) net charge. Up quarks and down quarks are similarly configured with the down quark having a little more mass. As we saw in our discussion on quark configuration, it is relatively easy to create a neutron by adding an electron/neutrino couplet to an up quark.

Protons and neutrons may exchange Pions that allow an identity exchange, but that may not be the sole (*or real*) source of a nuclear force. The Tor Model recognizes the relationship between fields as the source of force. The fields of protons and neutrons are very much alike. The particles are configured similarly, and their mass differential is only .14 percent. Because the two particles are so much alike, their spin characteristics are alike, so they produce fields that are very much alike. Being alike but not exactly the same, the fields are attracted to each other and can exist in close proximity, but the difference between them precludes them from uniting. Consequently, the attraction between the fields provides a nuclear force that exceeds the proton repulsion force and holds the nucleus together as long as there are at least as many neutrons as protons in the nucleus.

Given that neutrons naturally decay into protons in a matter of minutes, it makes sense for neutrons to have evolved after the hydrogen atom evolved, which allowed neutrons to readily join a nucleus before decaying. Their initial creation therefore would not be limited to a fifteen-minute window of time. Once inside a nucleus, neutrons become quite stable. It makes no sense for the neutron to have evolved before the hydrogen atom because without the need for the neutron in atoms larger than hydrogen, the neutron would have no reason to evolve; there would have been no niche to drive its evolution.

8.4c The Atom

Once the four stable subatomic particles evolved, the evolutionary development of the atomic Level probably went smoothly but not necessarily quickly. A simple Tryk combination creating the electron was likely quite stable and naturally selected early. In keeping with the universe's natural tendency to be in equilibrium, thus charge neutral, to be instrumental in the next Level of matter the electron's negative -1 charge needed an offsetting positive +1 charge particle. The anti-electron (*positron*) has a positive +1 charge value, but because it is the mirror image of the electron they annihilate, making it unsuitable for the off-setting task. The CAGI would have to develop different combination of quarks to do the job.

Although many Tryk combinations would have been created, natural selection chose the up quark configuration with a $+2/3^{rds}$ charge and down quark configuration with a $-1/3^{rd}$ charge to dominate. The most stable configuration was in threes with their partial charges adding up to the +1 charge (*+2/3, +2/3, -1/3*) of a proton, equaling perfectly and offsetting the -1 charge of the electron. Those configurations eventually allowed the hydrogen atom to form. But how did the universe know it needed a +1, positively charged particle?

It was not a conscious need but a need for cosmic evolution to progress. The CAGI compels matter to combine until it creates a new Level of matter with its own capacity to combine. Time is of no concern. The CAGI will take as long as necessary and create as many trial combinations as necessary for a combination to come about that qualifies as, or is a sub-stage to, the next stable Level of matter.

In this case, quark-like combinations and electron-like combinations would have been created over and over until configurations capable of combining into a larger, stable particle came about. Since the electric force was now available to bring particles together, it makes sense that the primary attribute driving the evolution of these particles was the attraction of their opposite spinning Tors and Tor-Chains. Once the stable electron with a -1 negative charge became dominant, it was only a matter of time before a Tryk combination came about to precisely off-set that charge value and create the proton. Once the composite proton came about, it could combine with the electron, creating the hydrogen atom.

From there, neutrons would evolve by simply adding an electron/neutrino couplet to the configuration of the proton. Neutrons provided a means for multiple protons to co-exist within a nucleus allowing for complex atoms to evolve. Protons and neutrons easily come together due to the attraction of their similar fields. A single proton and neutron combine to form a deuterium nucleus, aka heavy hydrogen. From there, various combinations can be combined and nucleuses created. That would allow two of those pairs to come together to create a helium nucleus. The conditions of the early universe could have allowed such combinations, resulting in a ratio of 75% hydrogen and 25% helium and a trace of others. No other particle combinations were necessary to create the initial atomic Level of matter.

The atom brought an exciting emergent quality to matter. Looking at any single atom, one would not expect it to have any special attributes, but when combined with other atoms, interesting things happen. One would not expect the simple combination of two hydrogen atoms and one oxygen atom to produce anything of interest, but in bulk it produces one of the most important elements in our universe – water. As you know, various combinations of atoms produce all the other elements in the form of gases, liquids, minerals, and metals that comprise us and our environment. Atomic combinations are truly wonderous emergent qualities, yet they pale in comparison to the emergent qualities that evolve in future Levels of matter.

The next evolutionary stage would be when stars created from hydrogen/helium gas swirled into existence awaiting SMBH maturity to pull them into galaxy formations, bringing us full circle from Chapter Three.

8.4d The Sequence of Particle Evolution

All particles are created from Tors.
Individual Tors → Tor-Chains
Tor Chains + Individual Tors → Tryks & Photons
Tryk-Chains → Electrons & Quarks & Neutrinos
Variously Charged Quarks → Protons
Proton + Electron + Neutrino → Neutrons
Protons + Neutrons + Electrons → Atoms

Again, the configurations of photons, neutrinos, quarks, and electrons are put forth only as an example of how those composite particles might be configured. The only certain attribute is that they are all comprised of a single early elementary particle like the Tor. In our exemplar configurations all particles are comprised of Tor and Tryk-Chains that can easily be broken down and quickly reconfigured, which accounts for how quarks, leptons, and particle pairs can be created from and decay into each other.

Because the universe was relatively large before particles were created, the mass-energy and temperature scale of the early universe was high, but not nearly as high as prescribed by the Standard Model. The combining and building in complexity of all these particles likely transpired under naturally occurring conditions and forces, without the need for ultra-high temperature, though at times the early universe may have been very hot.

We have covered the expanse of the BBT's first 380,000 years, when atoms were created and photons were set free. Let's look at how the observational evidence in support of the BBT worked under the Tor Model.

8.5 Observational Evidence for Big Bang Theory

The purpose of this treatise is to show there are alternative theories for the early universe that encompass the observational evidence said to support the BBT. We are at a point in that evolution to evaluate whether that hypothesis is holding true. Taken from Astronomer Carolyn Devereux's book, *Cosmological Clues,* there are three key pieces of observational evidence for the BBT. [7.]

1. The universe is expanding. From this observation one can unwind the clock back and argue that the universe had a beginning and was once much smaller and denser. The Tor Model's view of the early universe agrees with that observation, except that unwinding the clock back to 10^{-43} seconds may be an unnecessary extreme that assumes conditions that may not have existed. The BBT posits particles, fields, and forces into existence. A timeline that posits things and conditions into existence without a causal connection to a pre-existing state invites problems that are compounded by positing hypothetical cures. Any initial assumptions other than the universe started from a simple expanding speck of pure energy may not have been valid, as discussed in Section 2.1a. But, yes, there is little doubt the universe is expanding.

2. The ratio of hydrogen to helium in the universe is 3:1. The creation of nuclei and atoms in the Tor Model happened much later than the period prescribed by the Standard Model, but eventually very similar post-particle-creation environments could have evolved. Once the temperature fell below 3000K, there is no reason to believe the creation of hydrogen, helium, and traces of lighter elements was not the same in both models.

The sequence of events in the Tor Model suggests the universe is older than the 13.8 billion years currently estimated. That estimation is based on the Hubble expansion constant of 72 km/s/Mpc.

There is also a reasonable basis for calculations indicating the constant could be 67 km/s/Mpc, which would make the universe close to 15 billion years old, providing time for the Tor Model's creation of SMBHs to have taken place. The disagreement between the expansion values is known as the Hubble tension and is a hotly debated issue scientists are currently working on.

3. The detection of the CMB. According to the Standard Model, when the early universe eventually cooled to 3000K, it allowed electrons and protons to combine to form atoms, which freed photons that had been scattering off those particles. This sudden liberation of photons sent them soaring throughout the expanding medium and most of those photons are still flying untouched.

The Standard Model posits that the expansion of the universe stretched the wavelengths of that radiation causing it after 13.8 billion years to lose energy and cool down to just a few degrees above zero. The remnants of that early phenomenon can be seen today by the presence of a background of very cool 2.75K microwave photons that uniformly exist in every corner of the universe. Those remnants are called the *Cosmic Microwave Background* or the *CMB*. The uniformity of the temperature supposedly existed when the photons began their journey and those micro fluctuations in that temperature are the seeds for the creation of our galactic structure. The presence of the CMB is one of the cornerstones of the big bang theory and one of the yardsticks for measuring both the age of the universe and the rate of its expansion.

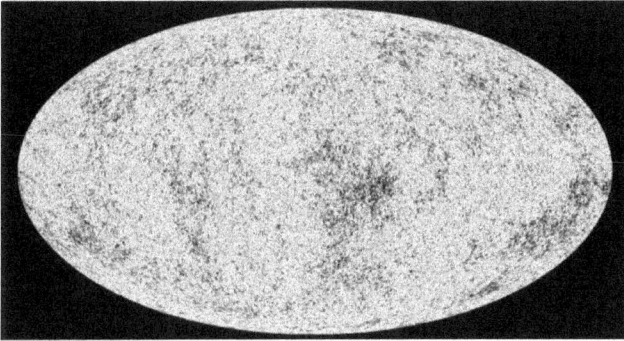

Figure 8.5 The Cosmic Microwave Background; provided by coolcosmos.ipac.caltech.edu. The light and dark areas show a very slight temperature difference hypothetically due to fluctuations in the conditions of the primal universe.

The Tor Model agrees with the notion that those photons may possibly be remnants of the early universe that were liberated when atoms formed. But the Standard Model's hypothesis for how the CMB photons lost their energy seems problematic if the interpretation of the double-slit experiment discussed in Chapter One proves particles to be real. If photons are real particles and not simply EM waves, then it is likely the forces holding each particle together, no matter how configured, are stronger than the force expanding the universe, thus expanding a photon particle and its wave signature through the expansion of the universe doesn't seem feasible. Also, the Standard Model's explanation implies that all those photons started the journey with nearly the exact same energy level, which seems difficult to believe. There must be another reason those CMB photons lost energy and arrived today at the same temperature.

Toward the end of the Cascade Effect when photons were first created, the universe was much smaller, homogeneous, and dominated by developing SMBHs. Though the SMBHs may not have been fully formed, they likely had sufficient gravitational influence to cause the photons to lose energy climbing out of the surrounding gravitational fields, a drag they would have endured for a large part of their early journey. This would have drained much of their energy, leaving a near uniform value with some variations.

Gravity curves space-time, which can change a photon's directional acceleration. Gravity does not change a photon's speed, which is constant at 3.0×10^8 m/s. Because a photon is massless and travels always at the same speed, it would seem that gravity has no direct impact on a photon except for its directional movement and therefore may not be responsible for its loss of energy. Gravity could be responsible for the photon losing energy, but for reasons not related to space time curvature but to gravity's influence on time. How a photon is influenced by gravity and time is discussed on Appendix #4.

In conclusion, although the timing of events would have been different between the two Models, the observational evidence and other support for the BBT is not inconsistent with the tenets of the Tor Model.

8.6 Conclusion to Part II

The source of force being the exchange of virtual bosons may be possible and make sense mathematically, but the required chain of hypotheticals, and the many other arguments against the notion found in Section 5.1, brings that scheme into doubt. If the IPM, particles, and their personal fields are real, then the dynamics of personal fields could be the source of attractive and repulsive relationships creating force. So, spin is the source of all fields and the relationship between fields the source of all force. (*Particle spin being the source of the gravitational field will be discussed in Appendix #3.*) Again, the benefit of simple relationships creating fields, forces, and composites is that it answers many outstanding questions using fewer hypotheticals. Most importantly, it provides a plausible evolutionary sequence to each of these developments.

As mentioned at the start of Chapter Four, it is with reluctance that I offer the possible presence of a core particle – the Tor – without proof for its existence. From an evolutionary standpoint it makes sense for there to have spun into existence a particle that proved sustainable and dominate, and from which all other particles are configured. Consequently, I was compelled to offer possible configurations of our Sub-A's with the proviso those configurations are only meant as possible examples and not precise structures. There is no direct evidence for our Sub-A's being composites, however, I'm confident that someday more compelling theoretical support for that notion will be developed.

If the outcome of the Double-Slit experiment proves particles with personal fields are real, it will weigh heavily toward the validity of a model like the Tor Model that provides a clear path from pure energy to particles, to fields, to forces, and to the Combination and Growth Imperative responsible for the evolutionary growth of complexity.

Next, we examine the prospects for a new reality should the outcome of the suggested double-slit experiment be as predicted proving particles and personal fields to be real. Such an outcome would undermine much of Quantum Theory's view of reality. How would a new classical view shape our perception of reality, ourselves, and our future?

Resolution of
Cosmological Issues – Ch.8

☒ Explanation of Work Energy
☒ Foundation of Quantum Mechanics
☒ Magnetic Monopole Mystery
☒ Missing Anti-Matter Mystery
☒ Neutrino Oscillation
☒ Origin of Cosmic Web
☒ Origin of Dark Energy
☐ Origin of Dark Matter
☒ Origin of EM field
☒ Origin of EM Waves
☒ Origin of Fields
☒ Origin of First Stars
☒ Origin of Force
☐ Origin of Gravity
☐ Origin of Mass
☒ Particle Decay
☒ Particle Evolution Sequence
☒ Particle Tunneling
☒ Purpose of Entropy
☒ SMBH Creation
☐ Space & Time Relativity
☐ The Fine-Tuning Problem
☒ The Flatness Problem
☐ The Hierarchy Problem
☒ The Horizon Problem
☒ The Magnetic Field
☐ The Measurement Problem
☐ The Reality Problem
☒ Virtual Particles
☒ Wave-Particle Duality Resolution
☒ What Drives Cosmic Evolution
☒ What is Energy
☐ Why Gravity so Weak
☐ Why Photons are Massless
☒ Why Three Generations

Part III

Bringing the Puzzle into Focus

Quantum Theory's interpretation of the delayed-choice experiment clearly makes QT appear weird, but from the Tor Model's point of view, the outcome is easily explained and not weird at all.

Chapter Nine
9.0 The Reality Puzzle

We have created a sound theoretical foundation to our early universe. The new vision accounts for the creation of SMBHs, the Cosmic Web, particles, fields, forces, and possible subatomic particle configurations. We may never be able to prove the existence of Penergy, pure energy blackholes, or early elementary particles, but their theoretical existence answers many cosmological questions and seems quite reasonable given the observational evidence and their theoretical simplicity.

We shall now evaluate whether we are appropriately interpreting this new vision. Stepping back and taking in the breadth of our puzzle we see billions of spinning, pure-energy, supermassive blackholes, proto stars, and very tiny, early elementary particles. The particles have combined and grown in Levels, building in complexity. Revealed at each Level were previously unseen emergent qualities that have included wondrous attributes such as fields, forces, photon absorption, and a variety of atomic combinations that will be used to build our universe from here.

That is what we see on the surface, but is there more going on beneath the surface? Could the presence, behavior, and interaction between all these components be playing out in accord with a deeper reality? Can today's Standard Models and the equations of General Relativity (GR) and Quantum Theory (QT) be the path to discovering a deeper reality? According to Physicist Lee Smolin, physics should be more than a set of formulas that predict what we will observe in an experiment; it should give us a picture of what reality *is*.[1] The discoveries and theories of particle physics and cosmology have given us road signs pointing to a subsurface reality. Let's head down that road to reality and see where it takes us.

9.1 Finding Reality

Let's start with a caveat: we are limiting our definition of reality to what is happening in the universe independent of a life-form or conscious observer. As Einstein once said, "Physics is an attempt to grasp reality as it is, independent of its being observed."[2] This point may be in conflict with some points of Quantum Theory, but they will be resolved in due course.

For many, reality is ultimately described in equations. So far, matter, energy, fields, forces, and their movements and relationships have lent themselves to being described quite accurately in mathematical terms. That success says a great deal about human curiosity and ingenuity. We seek to understand our world. We peer into the depths of space and the Levels of matter in hopes of learning the laws of nature that will give us a perspective on who we are, where we have been, and where we are going. We are a part of a complex nature, but one that can apparently be understood. There is much we know about it, but much we don't know. Given the success of our understanding so far and the success of our describing it mathematically, many believe there must be a description of a single reality that encompasses all there is.

The Standard Model of Particle Physics and the Standard Model of Cosmology are touted as being two of the greatest accomplishments of the twentieth century. Given what they encompass and what they have accomplished, they are indeed monuments to human endeavor. The mathematics supporting those Standard Models does not give us a complete picture, however. Physicist Frank Wilczek tells us the Standard Model of Particle Physics is flawed; its equations are lopsided and contain several loosely connected pieces.[3.] These models have provided us with a reasonable picture of the universe and how it works, but there are missing pieces to the picture (*dark energy and dark matter*), proven theories in conflict (*GR & QT*), and explanations that challenge credulity (*inflation and missing anti-matter*). These two models, as wonderful as they are, cannot yet provide us with a complete picture of reality.

General Relativity does a wonderful job of describing space, time, gravity, and the larger aspects of the universe. It predicted blackholes, gravitational waves, and time dilation, all of which have been confirmed many times. Its equations, however, only address those larger aspects of the universe and cannot be depended upon to describe the micro-world of quantum particles. By itself, GR is not likely the source for a complete picture of reality, but its success certainly suggests it will have to be accounted for in such a vision.

The mathematics of Quantum Theory accurately describes particles, fields, and the smallest aspects of the universe, but there is much controversy over whether QT represents an underlying reality. Physicist Niels Bohr and other architects of quantum mechanics believed it to be a good tool for describing/measuring aspects of particles, but they did not believe it should be taken as a broader theory as to what is real.

Einstein was uncomfortable with quantum theory, and over the years many others have felt discomfort in the disparity between the world as we see it and the accepted model of the quantum world where nothing is really what it seems. Physicist Michio Kaku tells us, 'Quantum Theory is the most ridiculous theory ever proposed in the history of science.'[4.] But, as he points out, it works and has withstood every experimental test hurled at it. Still, it is difficult to believe that a theory believed to be the most ridiculous theory ever proposed would at the same time form a sound basis for describing reality. QT accurately describes our micro world mathematically, but its inter-pretation of that world leads to some unusual conclusions, often referred to as quantum weirdness. Can such weirdness serve as a basis for our reality? We need to examine Quantum Theory.

9.2 Quantum Theory

The development of QT is an interesting story that starts in the seventeenth century around the time of Isaac Newton. Robert Hooke and Christian Huygens developed a wave theory of light, while Newton had developed a 'corpuscular' theory of light, both of which contained reasonable evidence for their respective theories. Later developments confirmed light to theoretically have both wave and corpuscular characteristics. With the experimental confirmation of Einstein's 1905 paper on the photoelectric effect, light energy was proven to be particles (*photons*), thus giving them definite particle-like *and* wave-like characteristics. Physicist Louis de Broglie later advanced a theory that *all* particles have wave-like attributes, which proved accurate. The recognition that two related particle attributes could not be precisely measured simultaneously gave rise to the

Heisenberg Uncertainty Principle and *superposition*, wherein a quantum object could be in two states at the same time.

All these observations and developments resulted in the need for a new mathematical structure that could explain the experimental outcomes of quantum particles. With the development of Schrodinger's equation containing a *wave-function* to represent the wave-like behavior of particles, Quantum Mechanics was soon born. All this brilliant work was accomplished nearly simultaneously before 1930. It has been refined and extrapolated since then, for the most part making the math better, but its interpretation even weirder.

There were different schools of thought on how to interpret quantum mechanics. One school was to make no interpretation at all: *the mathematics of quantum mechanics works very well so don't bother with interpreting its meaning.* That school of thought was known as the Copenhagen Interpretation, which was apparently influenced by Logical Positivism, a popular philosophy at the time. That philosophy was founded on the idea that the only things we can know are what we observe, and what we are not observing does not exist, or at least asking questions about it is senseless. The idea that particles don't exist if we are not observing them, and don't have attributes until we measure them, is known as a non-realist philosophy.

Einstein was a realist. He believed particles known to exist continue to exist whether observed or not, and that particle attributes exist both before and after measurement. For many years he carried on verbal battles with Niels Bohr who of course believed that any discussion about unobserved particles and their attributes was sense-less. Bohr was perfectly content with both quantum duality and quantum uncertainty and did not believe an underlying meaning was necessary. His philosophy was summarized with the expression, 'Shut up and calculate.'

An attempt to put quantum mechanics back into the 'realist' philosophy camp was put forth by Louis de Broglie, and later by Physicist David Bohm, in the notion of the *pilot-wave theory* discussed in Chapter One, but it was never a widely accepted theory. Another effort to put quantum theory in the realist camp was put forth by Physicist Hugh Everett, who suggested superposition was not an issue of certainty and uncertainty and that all states could be real if the universe *split* at each superposition. In this way both states are real, though in separate universes with different realities. This notion became known as the Many-Worlds Theory. The theory is considered unprovable and thus remains on the fringe of scientific theories.

Quantum mechanics has proven to be quite accurate and is therefore believed to represent a certain reality of nature. But it is fraught with multiple levels of quantum weirdness. Let's examine that quantum weirdness closer, starting with the idea of uncertainty.

9.2a Uncertainty

Science is often looked upon as the search for the truth about nature. For physicists, this largely means finding the truth about matter: how it moves and interacts, and how those things change with time. Newton's observations and mathematics described those things pretty well. But once scientists began exploring below the atomic level, the movement and interactions that changed over time were not so easily described by Newton's laws. The belief that particles could be both point-like objects and waves, and exist in two distinct states, created a duality difficult to explain, making the quantum world appear quite different from the macro-world we live in.

Quantum Theory deals in the realm of subatomic particles, many of which cannot be seen but are inferred from experimental evidence. Bouncing something as light as a photon off a particle to gain some information about the particle can change the particle's movement or characteristics. Even accounting for the bouncing effects, we can only obtain limited knowledge from a single encounter with the particle; if we determine its exact location, we cannot know its exact momentum and vice versa.

Looking deeper into the makeup of matter, we soon reached a limit to what we could observe or measure given the size and speed of subatomic particles. This sometimes leaves us uncertain when trying to obtain two characteristics of a particle simultaneously. That limitation is encapsulated in the *Heisenberg Uncertainty Principle* (HUP), developed by physicist Werner Heisenberg in the 1920s. It followed the development of Quantum Mechanics, the mathematical formulation that gives probable answers when dealing with particles and their interactions. Being so important to the foundation of quantum theory, we should take a closer look at the uncertainty principle.

9.2b The Heisenberg Uncertainty Principle (HUP)

In Quantum Theory, the HUP is a fundamental principle that quantifies the trade off when simultaneously measuring two quantum variables, such as position and momentum or energy and time. The principle was formulated when Heisenberg was trying to build an intuitive model of quantum physics. He discovered that there were certain fundamental factors that limited our actions in knowing certain quantities. He created a core equation that accurately expresses this limitation: the more we know about one variable, the less we can know about the other.

From QT's standpoint, for a particle to remain a 'particle' over time [*as opposed to a wave*], it must possess definite values of position and velocity, but the HUP denies this possibility, giving rise to a delocalized wave-like aspect of quantum particles.[5.] In other words, since a particle cannot possess both a specific position and momentum, it possesses a wide range of both values at the same time.[6] These theoretical ideas are a part of what makes the quantum world weird.

There are different points of view regarding the HUP. One is that while the HUP is an important and necessary principle that needs to be taken into consideration when predicting experimental outcomes, the principle is associated with a human endeavor and should not be applied to the description or behavior of particles independent of that endeavor. As Physicist Lee Smolin tells us, "...the terms by which science describes reality cannot involve in any essential way what we choose to measure or not measure."[7] From this line of thinking the appropriate interpretation of the HUP is that it is a very helpful mathematical deduction that simply expresses a human limitation in simultaneously measuring two self-created particle attributes. And, our inability to measure should not justify assigning to, or taking away from, any characteristics of the particles.

There are, of course, opposing opinions. Some physicists have adopted the HUP as representative of a fundamental aspect of particles. If the position of a moving object is known but not its velocity, in the next instant of time, having a range of velocities available to it, the object would effectively come to occupy a range of different positions simultaneously. In other words, it would become a wave of superimposed position states.[8] This interpretation says that objects subject to the HUP no longer occupy a specific position when their velocities are known and have no specific velocity when their position is known. When we know a particle's position, then it behaves like a particle. But it does so only instantaneously. As soon as we look away, the HUP dictates that the particle must start behaving like a wave again because its velocity is not well determined.[9] This idea of multiple states is called *superposition*.

Reason would lean toward the former explanation: the HUP simply expresses a human limitation and says nothing about the attributes of particles themselves. The limitation inherent in the HUP makes sense when one realizes two things. Firstly, one of the two quantum variables under consideration is in constant change. Momentum means the particle under consideration is moving; its position is constantly changing. Secondly, one must realize that there is no such thing as "now". Now is a term we use generally to refer to that point in time between past and future; it falls just after the past and just before the future. The nature of time precludes us from ever experiencing or measuring a precise "now" because before one can recognize and record it, it has already passed. Consequently, if a particle is constantly changing position due to it experiencing momentum, and there is no way to record "now" during that movement, there is no way to know with absolute precision the particle's momentum *and* position simultaneously.

One could pick a future time and say that if conditions don't change, meaning the rate of change remains constant, then one could *predict* a measurement value. This is in essence what quantum mechanics does; it predicts a value given a calculable rate of change represented in the wave function.

9.2c The Double-Slit Experiment

The double-slit experiment, detailed in Chapter One, has been at the core of Quantum Theory's belief that the subatomic world is both weird and uncertain. In QT, the photon (*or its wave-function*), passes through both slits. Quantum Mechanics predicts this result because it describes each individual photon as a propagating wave function that fragments at the slits, spreads, and intermingles with itself on the other side, creating a pattern of high and low probabilities for where on the detection screen the photon will land.[10.]

Placing a detector within the experiment to determine which slit the photon passes causes the interference pattern to disappear.

Physicist John Wheeler proposed a clever experiment called the *delayed choice* experiment in which detectors were set up *behind* the detection screen which was fashioned as a set of blinds that could be quickly opened and closed. The idea was to see if the photons could consistently choose whether to be a particle or a wave depending on whether they were about to strike the blinds or pass through the blinds and strike a detector.

With the capacity to open and close the screen-blinds quickly, the experimenters could choose the status of the blinds after the photon passed the slits but before encountering the blinds. The experiment was carried out in 1984 at the University of Maryland using high tech equipment. The result was that the photons always got it right, even if the choice to turn the path-tracking detectors on or off is delayed until after a given photon had passed through the partition.[11.]

The interpretation of this positive experimental result was that the photons could choose whether to be a particle or a wave retrospectively, *after* passing the screen. According to Physicist Thomas Hertog, "The delayed-choice experiment illustrates vividly and strikingly that the process of observation in quantum mechanics induces a subtle form of teleology into physics, a backward-in-time component. "[12.] This interpretation adds to our growing body of quantum weirdness.

As argued in Chapter One, the double-slit experiment can also be interpreted as the particle's field passing through both slits and causing the interference pattern, making particles real objects, just as they are always observed. Photons passing through only one slit each time would also explain the outcome of the delayed choice experiment, since the interference pattern would always show up with the blinds closed and always set off only a single detector with the blinds open, no matter when the blinds were opened or closed. QT's interpretation of the delayed-choice experiment clearly makes QT appear weird, but from the Tor Model's point of view, the outcome is easily explained and not weird at all.

9.2d Particle Entanglement

Another demonstration of particle peculiarity is the topic of *entanglement*. If two alike particles originate from the same source or otherwise meet, they are said to be entangled, meaning they somehow have the means to communicate over significant distances without anything passing between them. Einstein called this phenomenon "spooky action at a distance".

A classic experiment demonstrating entanglement involves photon polarization, the term used to describe the direction a photon is waving as it travels. We can't tell what direction the wave is pointing until passing the photon through a polarizer set at a certain direction. If it passes through the polarizer, we know that its wave was pointing in the same direction at which the polarizer was set.

In the basic experiment, if two photons originate from the same source and are then sent out in opposite directions toward two polarizers, the photons will do the same thing in terms of passing through the equally set polarizers. If one photon goes through a polarizer, the other photon will do the same. If the first photon is blocked at the polarizer, the other photon will also be blocked.[13.] How does the second photon know at what angle the first photon was traveling in order to mimic its behavior? The presumption is that information passed between them just before the second polarizer was encountered. Similar experiments have been carried out using other particle traits such as the spin-up/down characteristics of electrons, and in all cases, it appears the electrons were somehow communicating with each other to statistically arrive at the correlation predicted by quantum mechanics.

Entanglement allowing communication between distant particles without anything passing between them is based on some assumptions. It assumes the physical systems (*particles*) do not have definite properties prior to being observed or measured[14.]; that the properties of a particle aren't fixed until we measure them[15.]; and that the polarization angle cannot be preset by the photons before starting out.[16.] These are big assumptions. They are based largely on the non-realist approach consistent with the previously mentioned Copenhagen Interpretation of Quantum Theory.

Einstein believed particles could not exchange information without something passing between them, a concept known as *locality*. Consequently, particles could coordinate their statistically matching properties when touching or when being created, but this implied they possessed additional, non-apparent properties known as *hidden variables*. A mathematical theorem created by physicist John Bell in 1964 based on statistics and the HUP is said to prove that quantum mechanics predicted stronger statistical correlations in the outcomes of certain measurements than any local theory possibly could, and that hidden variables cannot be experimentally verifiable.

Einstein's hidden variables were unobservable traits that could potentially explain the seemingly random outcomes of quantum mechanics by providing a more complete deterministic picture of reality. As we saw in Chapter One, the concept of personal fields explains QM quite well. Accordingly, no other traits or variables are necessary to explain QM, rendering the need for hidden variables for that purpose unnecessary.

The idea of entanglement is being used in the development of cryptography and quantum computing and many experiments point to it being very real. Yet perhaps there is more for us to understand. Undoubtedly scientists are finding two entangled photons miles apart having the expected properties, but when and how they acquired those properties is still an open question. Hidden variables may not exist in ubiquitous fields but they very well may exist in personal fields. It could be that when the two particles touched, they synchronized the variables in their respective personal fields. Fields are disturbances in the Penergy medium due to the peculiar aspects of a particle's spin, thus some traits of the particle could be imparted to its surrounding field. When two particle fields overlap, they could absorb some of each other's field traits and begin sharing those characteristics. Fields would then be the only trait necessary to convey like characteristics, rendering Bell's theorem moot, or possibly even wrong due to wrong assumptions.

Entanglement could turn out to be a good argument for both particles producing personal fields and particles communicating through those fields. More analysis is necessary, and we must ensure we have a full understanding of personal fields and all their aspects and potentialities before drawing final conclusions. It is possible that our thin Penergy medium is a very dynamic form of energy capable of much more than we currently know. It could be the means for a common inter-connectedness between all things and the possible means for future cosmic communication. It is my hope that the scientific investigation into quantum entanglement will lead to suggestions as to how we might someday tap into that energy.

Notwithstanding all this quantum weirdness, quantum mechanics works very well. It consistently and accurately predicts experimental outcomes, though those predictions are expressed as probabilities. The original equation accounting for this ability is the Schrodinger equation containing the *wavefunction*. Let's examine the idea of the wavefunction and its effect on reality.

9.2e The Wavefunction

One of the issues in interpreting the meaning of quantum mechanics was a controversy regarding the meaning of the Schrodinger equation and its *wavefunction*. It entails the notion the particle under consideration can behave like a wave or a particle, with its measurement potentially having an infinite number of values that are evenly spread out (*much like a field was described in Chapter One*). When a measurement is taken, the wavefunction is said to collapse, producing the probability value for the attribute in question.[17] The uncertainty about the meaning of the wavefunction is expressed in a controversy over interpreting when exactly the wavefunction collapses, whether it requires a conscious observer, or even if it collapses at all.[18] That issue is known as the *measurement problem.*

The wavefunction is a mathematical term, not a real thing. No experiment has ever proven the existence of a wavefunction.[19] Quantum mechanics is accurate and is used widely throughout science and industry, consequently many see the wavefunction as something not only very meaningful, but very real. It has been elevated to the level of being part of reality, and for some it is, 'the sum total of reality'[20] In Quantum Theory everything has a wavefunction; your book, the earth, even the whole universe has a wavefunction.[21] An object's wavefunction determines its behavior and the behavior of an object's wavefunction is determined in turn by the Schrodinger equation.[22]

Again, quantum mechanics is capable of giving accurate probability answers, but perhaps its success may not be so mysterious. The experimental outcome determining a particle's attribute takes place when a measurement is taken and the wavefunction collapses. In the Tor Model, much of the wavefunction's success and accuracy can be ascribed to the wavefunction mimicking the particle's personal wavy field. To measure a particle's attribute, one must interfere with the particle, which collapses its field. Perhaps when the wavefunction collapses to reveal the value of a particle's physical attribute, so does the field around the particle collapse, exposing the particle's attribute. The collapse of the wave function approximately rhyming with the collapse of the particle's field may be what makes Quantum Mechanics seemingly so statistically accurate.

We have examined quantum uncertainty and superposition, wave-particle duality, quantum entanglement, and the wave-function of quantum mechanics. It is time for an assessment of whether quantum theory can be used as a basis for reality.

9.3 The Reality of Quantum Theory

Reality is the state of things as they actually exist, as opposed to an idealistic or purely mathematical idea of them. We once thought reality was just as we saw the world, but we are told that what we see is only on the surface of a deeper reality, a deeper existence. That deeper existence is described by QT in terms of such things as a dual nature of particles, the possibility nothing exists but fields, or the possibility we live in multiple worlds. All these notions suggest a reality much different than the one we witness daily. The enigma at the heart of quantum reality can be summed up in the expression from Physicist Sean Carroll: "What we see when we look at the world seems to be fundamentally different from what actually is."[23.] Given the weirdness of these quantum ideas, can Quantum Theory provide us with a valid picture of reality?

There is no question whether QT math works. The only question is whether the *interpretation* of the math represents the underlying reality of how the universe works. Some of the weirdness of QT's picture of reality is self-induced. It has adopted extrapolations of the HUP, such as *Quantum Mechanics cannot ascribe attributes until those attributes are measured.* Physicist Yasunori Nomura characterizes this assertion as a *metaphysical* claim.[24.] That means that it is not scientifically proven. It seems to be a concept left over from the Copenhagen Interpretation that was at the heart of the controversy between Niels Bohr and Einstein. Einstein of course believed that particles do have definite values for their position and momentum, and just because we can't determine them simultaneously doesn't mean they don't exist or that measuring only one allows the other to take on numerous values.

Another quantum theory extrapolation of the HUP is that if you know an electron's momentum you cannot know its location, and if you cannot know exactly where it is then it exists in several *parallel states simultaneously.*[25.] The Tor Model's answer to this is that you don't know the electron's precise location because it is flitting about within its bubble-like field. In Quantum Theory the electron can be flitting about anywhere in the universe, which is an extrapolation gone too far, creating its own weirdness.

Einstein believed that quantum theory, though giving correct results, was incomplete and he did not believe a wavefunction could be used to describe reality.[26.] Other physicists accept Quantum Theory as very much describing reality. They have extrapolated the math of the HUP and quantum mechanics into a broad spectrum of ideas including time does not exist[27.], nothing is real[28.], very little can be known with certainty, and we probably exist in one of many worlds with similar but different histories.[29.] An excellent account of all these theories can be found in Sean Carroll's, *Something Deeply Hidden.*

These ideas are principally math driven. Although math can produce accurate experimental predictions, it can be unreliable for describing reality. Tim James tells us quantum mechanics is a triumph of mathematical beauty, provided you don't ask what any of it means.[30.] Physicist Sean Carroll tells us that the things math proves are not true facts about the world[31.], and that physicists don't know what Quantum Theory actually *is*.[32.] The elements of Quantum Theory, though accurate mathematically, may not produce a usable picture of reality for the following reasons.

- Entanglement is a very interesting phenomenon, but it may be no more than that. Experiments have proven that particles have attributes, some of which are fixed and some that are subject to variation such as the direction of spin or polarization. Experiments have also proven that two particles, when entangled by touching or having emanated from the same source, can share those attributes in statistically predictable ways. Quantum Theory says that statistical predictability is due to the particles sharing the same wave-function and having the capacity to pass information rapidly that allows the second particle to adjust its attribute depending on the value of the first.

  When and how each particle acquired the value of the attribute and whether they can pass on or preset those variables is still an open question. Particles relating attributes without anything passing between them is a respectable theory, but so is personal fields being the source of shared traits that renders Bell's Theorem moot. Entangled particles are real, but how and when they acquired their entangled attributes is inconclusive, so the theory cannot yet be trusted as part of our reality.

- Wave-particle duality may not be real. If future experiments demonstrate that the waviness of a particle is simply inherent in the signature of its personal field and that it is the field that passes through the double-slit experiment causing the interference pattern, then the wave-particle duality issue may be put to rest. Wave-particle duality therefore would no longer be part of our reality.

- In Quantum Theory, if a particle's momentum is known, then its position can take on a variety of values, placing it in a state of a superposition of probabilities. If particles are real and not waves, it means they have definite positions, though difficult to pin down within the particle's personal field except statistically. It also means the idea of superpositions is dubious, and that a particle seemingly in multiple locations may be explained by it flitting about within its own personal bubble-like field. The superposition of particles is never actually observed. As Physicist Thomas Hertog tells us, "We don't observe a superposition of realities; experimenters find particles either here or there, not both here and there."[33.] Superposition may be only possible given wave-particle duality and consequently it is not a part of our reality.

- Schrodinger's wave-function equation provides accurate experimental predictions and measurements. The validity of the wave-function calculations, however, could be due to the wavefunction mimicking the actual wave-like attributes of the particle's personal field as explained above. The wave-function therefore remains accurate and very usable, but it is simply a mathematical tool to help us understand experimental outcomes and is not a description of a deeper reality.

- The HUP is an important concept and mathematical tool but is perhaps nothing more. Without the concepts of wave-particle duality and superposition, the HUP becomes a mathematical expression of a human limitation to measurement and not otherwise a part of a deeper reality.

Wave-particle duality, the HUP, superposition, and the wave-function predicting accurate probabilities, have all been brilliant answers to real, head-scratching problems. Those answers, however, created a picture of quantum weirdness that may have led us down mathematical roads to realities that don't exist. A reality based on probability is fraught with opportunities for miss interpretation. Lee Smolin tells us that the use of probabilities is just for our convenience and the resulting uncertainties just an expression of our ignorance.[34.] Physicist Kenneth Ford points out that the use of probability in the larger world results when we don't have all the available facts.[35.] That might be the case for Quantum Theory as well; we just don't yet have the means to know everything that is going on in a particle's minute environment. Physicist Nomura tells us we may never really know what is going on in the quantum world.[36.]

If the outcome of the double-slit experiment proposed in Chapter One is as predicted, then particles are real, they have personal fields, and forces are likely derived from the relationships between those fields. It means our reality will come from a realist point of view as Einstein insisted, solving the reality problem. Without superposition, it means quantum systems are not in multiple states, solving the measurement problem. It means QT is an excellent mathematical description of certain behaviors of particles at the quantum level, but it's not the basis of a broader reality. It means GR is an excellent mathematical description of gravity and related topics, but that is all, and that combining those two theories will not give us a deeper picture of a reality. As Physicist Mark Jones tells us, QT and GR are both good theories, but they are not compatible, consequently there must be a deeper theory.[37.] In Chapter Ten we will explore what that deeper theory might look like.

Resolution of
Cosmological Problems – Ch.9

☒ Explanation of Work Energy
☒ Foundation of Quantum Mechanics
☒ Magnetic Monopole Mystery
☒ Missing Anti-Matter Mystery
☒ Neutrino Oscillation
☒ Origin of Cosmic Web
☒ Origin of Dark Energy
☐ Origin of Dark Matter
☒ Origin of EM field
☒ Origin of EM Waves
☒ Origin of Fields
☒ Origin of First Stars
☒ Origin of Force
☐ Origin of Gravity
☐ Origin of Mass
☒ Particle Decay
☒ Particle Evolution Sequence
☒ Particle Tunneling
☒ Purpose of Entropy
☒ SMBH Creation
☐ Space & Time Relativity
☐ The Fine-Tuning Problem
☒ The Flatness Problem
☐ The Hierarchy Problem
☒ The Horizon Problem
☒ The Magnetic Field
☒ The Measurement Problem
☒ The Reality Problem
☒ Virtual Particles
☒ Wave-Particle Duality Resolution
☒ What Drives Cosmic Evolution
☒ What is Energy
☐ Why Gravity is so Weak
☐ Why Photons are Massless
☒ Why Three Generations

It's not that the universe created the conditions specifically for 'life' to exist, it's that the CAGI simply exploited those tenuous conditions and made life from them.

Chapter Ten
10.0 The Big Picture

The existing cosmic picture puzzle is incomplete but nonetheless conveys a plausible story of the evolution of the universe. It contains a mixture of segments, some that are beautiful and fit together well, some that work well but require a bit of mathematical manipulation, and some that work well but produce a fuzzy image. There are several segments that fit together well but reflect images that don't fit in anywhere else, such as Supersymmetry, String Theory, and the Many Worlds Theory. There are hand-carved, hypothetical segments that fill a need but provide questionable images of reality and are hard to believe, such as our current theories for inflation, fields, forces, and the origin of mass. And there are several missing segments, such as images for dark matter (if it exists), dark energy, and the missing anti-matter.

Despite these issues and shortcomings, scientists have managed to hold this complex puzzle together with a mathematical glue that speaks to both their creativity and the flexibility of mathematics. Overall, the picture represents a plausible picture of cosmic history, but the support for some of those puzzle segments, especially those representing the very early universe, have been tweaked to fit mathematically but are otherwise rather tenuous. The picture falls short of being clear and cohesive. There are too many hypotheticals blurring the image, and many pieces only fit together tenuously. It has been evident for some time that we are clearly missing something important.

10.1 In Search of the Big Picture

Our new segments to the puzzle reflecting the early universe fit together well and seem comparatively more cohesive. With fewer hypotheticals, they present a clearer image and solve several cosmological mysteries. Yet the overall ramifications of these hypotheses is still to be determined. Assuming at least some of the hypotheses put forward have validity, where do we go from here? We understand the limitations the reality of these images can convey, but is there more the picture can tell us? Is there a perspective on the universe we are not seeing with clarity?

We seek a complete cosmic picture that answers all the questions and is simple, elegant, and fits together with mathematical precision. We call such an all-encompassing picture a Theory of Everything - a TOE. String Theory is a leading hopeful, but it requires ten or eleven dimensions, and there is no apparent way to prove the theory by experiment. Supersymmetry is another hopeful, but it requires a whole bunch of new particles that have not shown up yet in our particle colliders. Many believe a TOE might be within reach once combining Quantum Theory (QT) and General Relativity (GR) into a theory of *Quantum Gravity*, but the two theories appear incompatible. And as Physicist Lisa Randall points out, the combination of GR and QT into a single set of equations that describes particles and their interactions might be possible, but they won't describe reality.[1.]

GR is a field theory that tells us how a mass responds to the energy of a gravitational field. GR parallels Maxwell's equations that tell us how particles respond to the energy of an electromagnetic field.

Another important puzzle piece is Plank's equation that demonstrates the relationship between energy and a photon. Each of these theories is distinct and have their own constant of proportionality (*G, k, and h respectively*), which is the link necessary to associate the movement of energy with a field or photon. These accomplishments are significant segments to our puzzle because they tell us a great deal about our universe by explaining how energy works. All of these are important, but none individually or in concert appear capable of providing a TOE.

Another scientific endeavor has been the Grand Unified Theory – GUT, which is the goal of combing the four forces into a single parent force. That too has proven only partially successful so far. It only seems plausible at ultra-high energies and temperatures, an environment created only in the HBBT. That notion is inconsistent with the Tor model, which is more in the nature a cold BBT, a CBBT if you will. In the Tor model the forces of nature did not splinter off a hypothetical master force but simply evolved out of the conditions of the early universe.

From the perspective of the Tor Model, the path to either a TOE or GUT must begin with the understanding of two things: Penergy and spin. They are at the heart of everything in the universe. Penergy spin creates particles; particle spin creates fields; field spin creates attractive and repulsive forces; and forces allow particles to move, communicate, and combine, giving structure and complexity to everything else that exists. Gravity is often referred to as the most influential phenomena in the universe, but gravity is one step removed. Gravity depends on spin [*examined in Appendix#3*], making spin the most influential phenomena in the universe.

Understanding Penergy and spin, one seemingly knows how the universe works from a physical standpoint at its most basic level. As pointed out earlier, the major scientific accomplishments by Maxwell,

Einstein, and others have been the description and relationship between fields and forces, and the nexus between those accomplishments is Penergy and spin. Once the mathematics is developed to describe Penergy and its dynamics, we will better understand supermassive blackholes, particles, and the Penergy medium, which comprise the foundation to our universe. We may then have the basis to combine all those properties into a single mathematical concept.

Prediction: Someday a physicist or cosmologist will create the mathematics to support the idea of Penergy, its dynamics and its relationships, paving the way for the development of a workable TOE.

10.1a A Super TOE

Understanding all those relationships might give us enough substance to create a workable Theory of Everything, but it would not be THE Theory of Everything. That Super TOE must take into account more than simply how the universe works mechanically; it must additionally encompass the depth and potential of Penergy itself. Penergy has also produced non-physical things, such as mind, consciousness, and intelligence. Encompassing those things into a TOE would be a start, but only a beginning, as we don't yet know all the dynamics of Penergy.

Those three subjects are of course *emergent qualities* as previously defined. Recognizing emergent qualities tells us that Penergy is much more than a simple substance we call energy and the CAGI is much more than a simple concept driving evolution. Penergy and the CAGI together have brought the universe a long way in terms of evolved complexity in the form of living, intelligent, technology producing creatures. The universe's future surely involves much more evolved complexity, with new emergent qualities we can't now even imagine. A true Super TOE may not be possible for generations to come.

10.1b Reality Check

Does this mean that none of the monuments to human thought, exploration, ingenuity, discovery, and creativity have produced a realistic perspective on the universe? Have we missed seeing the forest through the trees? That might well be the case.

To develop a true perspective on the universe one must frame the picture accurately, starting with the dynamics of Penergy, the substance from which everything is made. Penergy and the changes it has undergone is not only the basis of cosmic history but cosmic reality as well. The dynamics of Penergy of course include the work of the CAGI, the ultimate source of cosmic change. Their work together can be encapsulated in a few lines, which I shall repeat for emphasis.

Penergy produces particles.
Particles produce fields.
Fields produce forces.
Forces produce Levels of complex matter and life.
Life produces intelligence and technology.
Intelligence & technology will produce greater complexity.
Greater complexity will produce even greater complexity.

Fig. 10.1 Cosmic Evolution expressed in terms of the CAGI

This is the substance and reality of the evolution of our universe thus far. There is more to come and more lines to add. The changes at each Level are becoming greater and the emergent qualities at each Level more wondrous. From this vantage point, the cosmic perspective is much bigger than the mechanical theories working at either the micro or macro level of existence. So, where does that leave us in our effort to find a meaningful picture? Reality is proving to be puzzling!

10.2 Cosmic Perspectives

A jigsaw puzzle is two dimensional, which is easy to describe mathematically. But the picture of our universe is not simply two dimensional. Mind, consciousness, and intelligence give it depth – a third dimension if you will. That depth comes from emergent qualities that are revealed at each Level of evolution. That depth may prove difficult to incorporate mathematically into a TOE. This third dimension is, however, a realistic element in the overall picture of cosmic reality and cannot be ignored. Apparently, the picture of cosmic reality expanded dynamically with the evolutionary development of consciousness and intelligence in the living organism. Consequently, to develop a complete picture of reality we need broader perspectives. Let's start with the concept of life.

10.2a The Life Perspective

Humans naturally think of life, especially human life, as something special. We are conscious, intelligent beings, with the extraordinary capability of understanding the universe. The creation of life out of inanimate materials is nothing short of miraculous. We have a soul. We know God. Of course we are special!

Perhaps we are. But perhaps we are simply complex organisms representing the earthbound Sixth Level of matter according to the Combination and Growth Imperative – the CAGI. The six Levels of matter are: 1st-Tors, 2nd-Tryks, 3rd-Sub A's, 4th-Atoms, 5th-Cells, 6th-Organisms. We are endowed with a level of consciousness and intelligence that no previous Level seems to possess. But these are simply emergent qualities inherent in our Level of evolution. Every Level has its own emergent qualities.

The Seventh Level will be created when we humans combine (*if we survive without annihilating ourselves*). That Level will have its own emergent qualities compelling them to feel superior. They will define their existence in a term more appropriate to their attributes, and so endowed they will believe they are "special". It sounds a bit egotistical, but that is a good thing. The belief in their superiority will keep future Levels working to better their lot, keeping them motivated to develop the resources and means to create the next Level.

The special existence of life supports the *anthropic principle* – the idea that the universe's parameters are set to allow life to exist. The essential elements in the universe - matter, energy, forces, and their relationships - can all be measured and reduced to about thirty numbers called universal constants. Viewing those constants, it has been noted that they appear to be fine-tuned for the existence of life and that if we change one of the constants by a relatively small amount, we find that it makes the universe inhospitable to life.[2] Why is the specifications for the universe so hospital to life? This is known as the *fine-tuning problem.* It is one of nature's deepest mysteries.

A belief that the universe was designed to suit life seems compelling. Stephen Hawking recognized the argument and after listing all the elements that had to be just right for us to exist, he observes, "Our universe and its laws appear to have a design that is both tailor-made and support us, and if we are to exist, leaves little room for alteration."[3.] Chemical Physicist Jim Baggott, however, reminds us we are very much a naturally evolved part of this reality, but we are not the reason for it.[4.]

In the big picture, we humans are merely a product of the universe due to evolutionary natural selection. It's not that the universe created the conditions specifically for 'life' to exist, it's that the CAGI simply exploited those tenuous conditions and made life from them.

There are over 100 billion galaxies, each with 100-200 billion stars, each with an untold number of planets and moons. If even a small fraction of them is suitable for the biological experiment of life, it means the odds are we are not alone in the universe. If the CAGI theory is valid it would mean there are Level Fives, Sixes, and perhaps Sevens or Eights throughout the universe and we are definitely not alone.

Alone or not, we are the trustees of planet earth, and it's quite possible we are also a rare form of intelligent life. As a global society we need to acquire a new image of this bigger picture of who we are and what our responsibility is. Given the devastating technology we hold in our hands, we can no longer afford to act like school-yard children squabbling over possessions, territory, rights, rules, what God said, or whose God is supreme. That behavior is just not acceptable.

As a global society, we need to adopt a new perspective on life, what it means, and how we are going to respect it. If we are indeed special, we must realize the responsibility that comes with that privilege. Our primary goal must be to survive. That means we must take care of our planet and ourselves, and plan for the future. Assuring our survival as a species will mean someday leaving our planet and building colonies in space and on other planets or moons. This feat is best accomplished as a global effort. Creating such a global image, uniting in the common goal of manifesting our destiny in space, and working together to create a larger, more complex organism, will certainly give us a new perspective on the concept of life. I believe everything discussed above is a part of our cosmic reality and will have to be accounted for in any Super TOE.

Our cosmic puzzle may even be four dimensional. The fourth dimension is, of course, *time*. Time comes into play because the puzzle is constantly changing. It is not only expanding incrementally but is changing in unforeseen ways driven by the CAGI. We need a perspective on this dimension of change.

10.2b The CAGI Perspective

Those who theorize the ultimate destiny of the universe seem to consistently limit that destiny to three options: the universe will expand to a point and then collapse into a *big crunch*; the universe will expand to a point and then slow to a crawl, expanding gradually but never actually stopping; and finally, the universe will expand forever, until all matter is so separated that the universe becomes a cold, dark, lifeless, abyss. Those theories are based on the physics relating to the geometry of space and the close balance between the strength of the energy expanding the universe in opposition to the strength of the gravitational energy contracting the universe. We never see a destiny theory that encompasses the richness of our cosmic evolution and how it may play a part in our cosmic future, influencing the fate of our universe. That richness has been driven by the CAGI.

The basic tenets of the CAGI are simple. It began with Penergy creating the first Level of matter by spinning particles into existence. At each ensuing Level, the dominant form of matter goes through the evolutionary process, combining in different configurations, with those configurations being tested by the hazards of its environment. This process continues until a stable form is created that is strong enough to dominate and is capable of combining. During each Level's development, emergent qualities evolve that assist in that Level's survival, development, and ability to combine. We humans seem to be the dominant Sixth Level here on earth. We possess the intelligence and technology to combine, and if we survive long enough, we are destined to do so.

How humankind could possibly come together to create a larger entity is not as difficult as one might think; it's a matter of technology. Reviewing the history of humankind, especially the last two hundred years, one understands how we are destined for that role and are currently very much on track to achieving it. Practically everything we have accomplished in recent history has been unknowingly oriented toward our combining into a larger entity. It is as if we are being guided by an unseen force, and we are globally doing exactly what we need to do to accomplish that combining process.

The story of that future evolution is told in my *The Journey of the Universe – 2nd Ed*. It describes in detail how we will combine and it looks into the future well past the Eighth Level, speculating on what cosmic evolution might be all about. It reveals how there may be much more in store for humankind and the universe than the typically pronounced endings of a big crunch or a slow drift to an abyss.

The story of Penergy has allowed us to peek inside the toolbox of the CAGI. Set against the backdrop of the Penergy medium, it appears the CAGI has needed only three tools to carry out its work: 1) Tor particles – the substance from which to build complex matter; 2) fields – that which creates *forces* and *work energy* that affects the interactions and relationships between them; and, 3) photons – transferring work energy to particles that provide them the means to couple/uncouple, form various phases (*i.e., gas, liquid, solid*), or otherwise exchange needed energy.

Like a fine painter meticulously applying color to her canvas, the CAGI molds matter into forms having surprising emergent qualities impossible to imagine given the apparent simplicity of the tools from which it must work. In much the way we appreciate a fine painter working with such simple tools, we are in awe of the CAGI's handiwork, and its contribution to the rich texture of our cosmic puzzle. It, too, will have to be accounted for in any Super TOE.

10.2c The Cosmic Perspective

It is apparent that a meaningful picture of reality is more than simply understanding the mechanics of how matter and energy work on a micro or macro level. It is more than understanding how the universe has evolved, or how life has evolved. Because the universe is forever growing in complexity in unanticipated ways due to the CAGI, a complete picture of a cosmic reality may not be known until we are closer to the manifestation of the universe's purpose and destiny.

It is also apparent that the destiny of the universe is far from ending in either a big crunch or a cold abyss. While much of our cosmic history has been structured by the persistence of the CAGI and refined by natural selection, there are far too many wondrous things evolving for this complex cosmic dance to have been choreographed by mere chance. The creation of emergent qualities tells us there is much more in store given the capacity of Penergy and creativity of the CAGI. The universe is surely on a journey. The most incredulous thing about it is that humankind may be in the envious position of being capable of understanding its mysteries, as well as possibly being a small link to its destiny. The persistence of the CAGI and its capacity for developing spectacular emergent qualities tells us there are fantastic experiences in our future. We need only to survive to become part of that cosmic destiny.

The evolutionary sequence appearing in Figure 10.1 has not yet been reduced to mathematical terms, but it seems to be a pattern that lends itself to such a formulation. Such a formulation would be looking at the universe from a new and different perspective; one with perhaps a sharper focus. Looking backward in time and deeper into the levels of matter does not seem to be giving us the picture of the cosmic reality we seek. We need to broaden our horizons, perhaps exploring a shift in focus. The reductionist philosophy that has existed in physics for the past century is not really getting us the vision we are seeking. Some physicists believe there is a quiet intellectual revolution taking place around the reductionist paradigm being false.[5.] Unpeeling the onion in the microworld won't give us a final theory, only the core of the onion. A final theory will come from envisioning how the outer layers of the onion are yet to grow. A change in focus from the past to the future, and from the depths of matter to the future of evolving matter may give us a better picture and a broader perspective.

The perspectives we have examined tell us there is a much broader vision of the cosmic puzzle than we have glimpsed thus far. The Tor Model has given us a vision of how the universe may have started and evolved, but more importantly it has given us an introduction to Penergy and the CAGI. These two things are essential to understanding the picture puzzle of the universe from any perspective.

Penergy is more than simply pure energy that creates and moves matter; it is the dynamic substance from which is molded the many unique qualities of our entire universe. Cosmologist Lawrence Krauss in summing dark energy says, "It is natural to suspect that its nature is tied in some basic way to the origin of the universe. And all signs suggest that it will determine the future of the universe as well."[6.] Dr. Krauss's suspicions are correct. Dark Energy (*Penergy*) has everything to do with the universe's unfolding, evolution, and destiny. Penergy is near magical, and for us the prospect of witnessing more of its exciting creations and surprises is exciting.

The CAGI is the force that molds Penergy's unique qualities, bringing into existence fields, forces, relationships, complexity, intelligence, and life. It will mold future combinations of matter into forms we can now only imagine. It will create complex emergent qualities of which we shall be in awe. The CAGI is the ultimate shaper of puzzle pieces.

Life is the Level of matter possessing the intelligence and technology to mold matter at an accelerated rate, transcending natural selection in its evolutionary process. Natural selection will remain a factor, but from the Sixth Level on, the Level of matter itself will be partially responsible for engineering its form, features, and future survivability. The definition of "life" will surely be redefined as new Levels of matter evolve.

In addition to being the descriptive language of the universe, mathematics will be the array of tools we will forever need for material development and problem solving to secure our survival. Mathematics can be beautiful and elegant, but at times be without meaning. Our challenge is to recognize when it is accurately describing reality and when we are reading more into the equations than they deserve. As Physicist Savine Hossenfelder tells us, "Physicists may not consciously subscribe to the idea that math is real and when asked will deny it, but in practice they do not distinguish the two. This conflation has consequences, for they sometimes erroneously come to think their math reveals more about reality than it possibly can."[7] Despite the dangers of occasionally leading us astray, mathematics is an essential element to our future survival and will have to have its own part in any future Super TOE.

10.3 Cosmic Reality

Given our understanding of Penergy, the CAGI, of Life, and Mathematics, what is our picture puzzle telling us about reality. Despite many missing, conflicting, and blurry puzzle pieces, we have a decent image of the universe from both a historical and mechanical standpoint, but what does the big picture really look like? It's complicated, and we don't yet have a usable TOE to give us a definitive perspective.

Personally, I believe the puzzle may be telling us there is more to the universe than its observable mechanical aspects and that cosmic reality is about *change*. Change implies looking toward the future. To get a glimpse of the true big picture, we must incorporate the CAGI and change into our vision. Matter isn't simply the manifestation of Penergy in the first second of the big bang, it is the dynamical seed of

Penergy that has been growing in complexity since its birth. The evolution of matter to life was no accident; it was an inevitable progression of complexity that is occurring throughout the universe.

From our many perspectives it is obvious that we are on a wonderous journey. We are a life form possessing the intelligence and technology to influence our evolutionary development, allowing us to someday move off our planet to secure our future. In hindsight, our technological development has been unknowingly pushing us in that direction. That technology is allowing us to modify ourselves in preparation for adapting to space and develop the physical means to live there.

In the backdrop of Penergy and the CAGI, we are products of an evolutionary journey that is inexorably building something very special. That something may be comprised of intelligence and technology, both of which we possess, creating the prospect of our playing a small part in that ultimate destiny. We know not the specifics of our immediate future, but our limited perspective suggests the overall course we must undertake to be a part of the cosmic destiny is for us to survive, to explore, and to evolve. The lessons we've learned from our different perspectives provide the knowledge, blueprint, direction, and means to accomplish that goal. We are not only sightseers on this wondrous journey; we are a part of the journey itself.

*　*　*

Thank you for joining me on this wondrous journey. It has been a wild ride and one that leaves us with a vision of the universe from a unique perspective. A perspective that only comes about if we are daring enough to view the universe not only from its mathematical description, but from an evolutionary basis derived from our observations, deductions, and imagination. It is hoped that this new vision provides a better foundation not only for understanding the journey of the universe, but the journey of humankind, as well.

It is a vision much different from today's accepted vision, but I have faith that some of the truth espoused herein will eventually seep into the scientific consciousness. As Physicist Luciano Rezzolla tells us with regard to an important lesson that history has taught us, "What appears exotic and obtuse today can become plausible and acceptable tomorrow."[8.]

Physicist Don Lincoln in his book, *Einstein's Dream,* points out that to find a TOE we will have to do it the old fashion way – by studying the world around us. He says, "Occasionally we will notice a loose thread in our theories and find that a tug on it unravels the entire fabric of our understanding of the laws of nature, and we will then use that loose thread to reweave it into a newer and more beautiful tapestry, one that better represents the way the universe actually works."[9.] Perhaps the concept of Penergy and the Tor Model espoused herein is that thread

Resolution of
Cosmological Issues – Ch.10

☒ Explanation of Work Energy
☒ Foundation of Quantum Mechanics
☒ Magnetic Monopole Mystery
☒ Missing Anti-Matter Mystery
☒ Neutrino Oscillation
☒ Origin of Cosmic Web
☒ Origin of Dark Energy
☐ Origin of Dark Matter
☒ Origin of EM field
☒ Origin of EM Waves
☒ Origin of Fields
☒ Origin of First Stars
☒ Origin of Force
☐ Origin of Gravity
☐ Origin of Mass
☒ Particle Decay
☒ Particle Evolution Sequence
☒ Particle Tunneling
☒ Purpose of Entropy
☒ SMBH Creation
☐ Space & Time Relativity
☒ The Fine-Tuning Problem
☒ The Flatness Problem
☐ The Hierarchy Problem
☒ The Horizon Problem
☒ The Magnetic Field
☒ The Measurement Problem
☒ The Reality Problem
☒ Virtual Particles
☒ Wave-Particle Duality Resolution
☒ What Drives Cosmic Evolution
☒ What is Energy
☐ Why Gravity is so Weak
☐ Why Photons are Massless
☒ Why Three Generations

Part IV

The Appendix of Puzzling Cosmological Issues

While developing the Tor Model I created many hypotheses, most of which fit easily into the model. There were some, however, that had merit, but I could not find good observational evidence for, though it may be found in future experiments or observations. They are concepts answering questions of why and how regarding important issues that can't be ignored. Consequently, I am including those notions outside the model and putting them in the form of appendices. Developing these speculations was part of the fun in my goal to create a new vision of the early universe.

Appendix #1

The Origin of Particle Mass

Penergy does not have *mass* as we define it until it is spun into a blackhole or particle, which then gives that Penergy content, definition, and location. Mass could simply be a measure of Penergy content, but how could there be massless particles like the photon? The photon could be comprised of so few, lightweight Tors as not to be measurable, but that would make the heavier particles comprised of far too many Tors, a complexity not likely to evolve so early in the universe. It could be that mass is not a measure of Penergy content, but something commensurate with it. It's a quandary. As Chemical Physicist Jim Baggott tells us, "I don't want to alarm you unduly, but no matter how real the concept of mass might seem, the truth is that we've never really understood it."[1.] Let's see how the Standard Model handles particle mass and examine a possible alternative theory.

The Standard Model's Theory

The Standard Model posits the existence of a special field called the *Higgs Field.* It is named after Physicist Peter Higgs who with others was instrumental in creating the theory for its existence. The field supposedly exists throughout the universe and gives mass to particles. In other words, it is the particle's interaction with this undetected, hypothetical field that causes particles to have mass.

The field is theoretically created in conjunction with the Higgs boson particle. It is a very large (*125,000 MeV*) particle that has not actually been seen but is believed to have been brought into existence at the Large Hadron Collider (*LHC*). Inside the LHC it is very unstable and survives for only 10^{-22} of a second.[2] That is less than a millionth of a millionth of a billionth of a second. The creation of the Higgs boson is a rare occurrence happening within the LHC only once in every ten billion particle collisions.[3]

Again, the Higgs boson has not actually been seen. Its existence is implied by interpreting the decay debris detected in the collider.[4] Reading debris created in particle colliders is notably imprecise. The process has been compared to throwing a piano out the window and then trying to determine all the piano's properties by analyzing the sound of the crash.[5]

The Higgs boson was initially predicted to be much larger. For the calculations to come within the vicinity of the mass observed, the mass number had to be *fine-tuned*.[6] Physicist Ben Still tells us that the lack of anticipated size of the Higgs Boson, "...is a big argument against the Standard Model and is known as the Hierarchy Problem."[7] Notwithstanding these problems, a predicted collision debris formula was observed in 2012 at the LHC, and the Higgs boson was reported discovered.

The Higgs boson supposedly maintains the Higgs field, yet the boson decays very quickly, so it does not naturally exist in our universe. Particles theoretically gain mass by interacting with the otherwise undetectable and ubiquitous Higgs field in a way that slows the movement of the particle, which causes it to act like our everyday understanding of inertial mass. As Physicist Lawrence Krauss tells it, if the Higgs field can make a particle more sluggish, the particle *acts* like it has a commensurate amount of mass.[8]

The problem with sluggishness equaling inertial mass is that it seems inconsistent with Einstein's E=MC². According to Einstein, the mass of a body is a measurement of its rest energy content. Apparently in the Higgs field, the particle does not actually possess the requisite rest energy content represented by its mass, it only *acts* like it does. Mathematics has been created to support the Higgs theory, but apparently there is no actual evidence the Higgs field exists, except by the presumption that if a particle exists then a corresponding field exists.[9.]

While the hypothetical Higgs is widely believed to exist, some physicists have reservations. Physicist Harald Fritzsch tells us that physicists invented the hypothetical Higgs field simply out of the need for a theory to give particles mass.[10.] Astrophysicist David Lindley tells us the Higgs field explanation for mass is ingenious, but it is a trick, a gadget, a kludge, as computer programmers would say of a piece of code that is tacked on to a piece of software to perform some necessary but overlooked task.[11.] Physicists Don Lincoln tells us, "…the proposed Higgs field did not arise from any fundamental principle. In a sense, one can think of the Higgs proposal as a band-aid that patched up a nasty wound in the electroweak theory."[12.] Lawrence Krauss tells us physicists don't have the slightest idea why the Higgs boson and field have the properties they do; "It seems completely ad hoc – a convenient addition to the theory that makes it work and allows the physical world to exist…without any underlying compelling mathematical rationale."[13.] Obviously, there are some serious concerns about the validity of the Higgs scenario.

The Higgs boson might be everything the physicists say it is, but producing a knot of medium-light-density Penergy by slamming protons together will always produce a big, unstable particle because that is how Penergy works. Predicting such a big particle will decay into smaller particles with a certain configuration is not convincing evidence for a particle producing an undetectable field that hypothetically gives particles mass. Slamming high energy particles together to produce a knot of Penergy and a cascade of lighter particles only supports my theory for the presence of our thin-density Penergy and the instability of heavy particles created within it.

The Higgs theory may satisfy a theoretical need, but it is difficult to believe and not foolproof, so particle mass could just as well be attributed to Penergy density or some other particle dynamic. I've always disliked the Higgs approach to mass and have pondered the issue at some length. If our Sub-A's are composite particles, then it makes sense for mass to simply be the sum of the masses of the constituent Tors making up the mass. Accordingly, mass is merely a function of particle configuration, which in turn is a function of natural selection given the dynamics of our frenetic, thin density Penergy medium. But that begs two questions; how can a particle be massless and why is mass in proportion to particle inertia? Let's speculate on a possible alternative hypothesis for mass.

An Alternative Hypothesis for Mass

Seemingly, mass could simply be a measure of its content in mass energy. In QFT, however, particles have no content, they are simply excitations of their respective fields. But, if we were to recognize particle content we would have to deal with the photon, which appears to have content but is massless. A photon is a particle, not a wave. It is not simply a bundle of energy, but is a composite particle made up of Tors. It can knock electrons off a sheet of metal, as shown in Einstein's photoelectric effect. Photons display a torque effect upon impact with a target. Photons possess a measurable momentum. All these things suggest photons have content, yet they are considered massless. Obviously, mass is not simply a direct measure of particle content. Consequently, we need a theory of mass that does not require a special kind of [Higgs] field, has some relationship with the particle's content, but can account for massless particles. So, if mass is not due to an interaction with a ubiquitous field, and not a direct measure of a particle's content, what is it?

Let's start with two associated observations regarding the nexus between mass, spin, and the spacetime-Penergy medium. Einstein taught us that mass curves spacetime. Fast spinning bodies such as planets and blackholes also curve spacetime.[14.] If *mass* can curve spacetime, and spinning bodies can curve spacetime, then perhaps what is actually curving spacetime isn't mass per se but something commensurate with mass, such as the spin of the mass itself or the spin of the particles making up that mass. It sounds crazy but this hypothesis works out to be very plausible. Under this notion, particle mass is much different than supermassive blackhole mass, which is examined in Appendix #3.

All matter is comprised of Tors, which are the only particles spinning on their own axes and being pure energy presumably do so at a very high rate. Things that spin on an axis at a high rate create a *gyroscopic effect*. That means that the spinning body is inclined to stay in the same position relative to its spin axis, resisting any tilt to that spin axis. You may recall this phenomenon if you have ever played with a toy gyroscope. Physicist Frank Wilczek describes elementary particles as ideal, frictionless gyroscopes that never run down. He also points out that the faster a gyro rotates, the more effectively it will resist attempts to change its orientation.[15.]

The measurement of a particle's mass refers not to the amount of matter the particle is made of, but rather to the particle's inertia, meaning the degree to which it resists being accelerated by a force.[16.] For a spinning body, that resistance to change refers to its axis orientation and is called *orientational inertia*, and the quantity that measures orientational inertia is called angular momentum.[17.] So, the more angular momentum a particle has, the more resistance there is to a change in its orientational inertia.

The spin axes of individual Tors within a composite particle point in various orientations, are not fixed, and are subject to subtle and continuous motion due to field dynamics and forces. Because the magnetic field connections between Tors are secure but not fixed, the fields are subject to minute wobbles and precessions. Additionally, because elementary particles contain both magnetic and electric fields, the Tors should experience an array of subtle movements as the fields of the particles constantly interplay. These dynamics would cause the precessions and forces on the Tor axes to be in subtle but continuous motion.

Unencumbered Tors are naturally driven in the direction of their spin axis. When in a fixed configuration with limited natural movement in the direction of its axis, Tors would have a strong resistance to perpendicular lateral movement. All those subtle motions would, however, manifest into a smooth and balanced equilibrium of motion that would allow the overall gyroscopic effect to take place evenly throughout the body, affecting all orientations. Any outside force on the particle affects that equilibrium and is resisted. All these dynamics add up to create an inertial mass that resists a change in movement.

It is seemingly a crazy idea, but perhaps a body's overall resistance to a change in speed or direction is simply the resistance inherent in its constituent Tors all acting like mini gyroscopes. The more spinning Tors that make up the body, even in all sorts of orientations, the more resistance to movement and hence the more measurable its mass. It is the reason mass is looked upon as a measure of a particle's content, but the existence of massless photons tells us that is not the case.

Massless Particles

Only in a situation where all the spinning constituent Tors are perfectly aligned would one find a massless particle. The only massless particles are Tors, Tryks, and photons. They are massless because all their constituent Tors and Tryks are configured to travel parallel to their axes of spin, as illustrated in Figure 1. Consequently, their movement does not necessitate a change in, or challenge to, their orientational inertia. Having no challenge to their orientational inertia, all three particles travel at maximum light speed. The fact they travel parallel to their spin axes is evidenced in photons having a torque effect on impart with a target, and Tor-Chains comprising a bar magnet showing a vortex effect at the end of each pole.

 ⇐ Direction of Travel and Spin axis	 ⇐ Direction of Travel
Fig. 1a. Tors, Tryks, and Photons have spin axes parallel to direction of travel and present no gyroscopic resistance. They are massless, propelled by their relationship with the Penergy medium.	Fig. 1.b Composite particles have spin axes non-parallel to direction of travel and resist movement in that direction. Combined with its gyroscopic resistance to tilt produces their inertial mass.

In conclusion, composite particles possess Tors and Tryks with various axes of orientation that possess an orientational inertia that resists movement consistent with the number of particles involved. That is why mass is ordinarily attributed to particle content. When all the spin axes are in a single alignment, the natural high-speed move-ment of the particle becomes a function of its special relationship with the Penergy medium. It's as if the parallel alignment with travel direc-tion gives the particle spin a propeller that drives it through the Penergy medium. If the spin axes are not in a parallel alignment, that special relationship does not exist, and the particle now resists any change in motion.

Binding Energy and Mass Defect

When particles join together, such as in the creation of a helium nucleus, the measured mass of the nucleus is slightly less than the combined masses of the individual protons and neutrons. This loss of mass is known as Mass Defect. Binding energy is a measure of the strength of field attraction when particles combine directly through their fields. It is equal to the energy it would take to separate the joined particles. The missing mass energy is measured to be the same as the energy binding the particles together, which is interpreted as the missing mass energy turning into kinetic energy that is passed along to the newly formed particle. In the Tor model, those energies are looked upon differently.

Particles at rest are never perfectly still and always possess some degree of jittery motion and kinetic energy. When the protons and neutrons come together to form a nucleus, what they give up is not mass energy but kinetic energy. When the nucleons are joined through their fields, they are then restricted in the degree they can freely jitter about; they give up some of their motion energy in the way of photon energy. The motion energy is in proportion to the strength of the field energy attracting and pulling them together and restricting their movement, making the mass defect and binding energy to have the same value.

More on Mass

This alternative theory for mass makes sense, but there is perhaps still more to mass than we currently know. For instance, the mass of a proton is measured to be about 938 MeV, whereas the mass of each of the three quarks supposedly making up the proton is estimated to be less than 5 MeV. There are many theories for how this could be, but none of them is compelling. Likewise, the mass of a Pion is measured to be about 140 MeV, but it is comprised of only two quarks each with a mass of less than 5 MeV.

Perhaps it is the state of the quarks that makes it difficult to determine their mass. If quarks are part of a long chain of particles, their freedom of movement would be limited and determining their mass a challenge. Perhaps it is the level of physical freedom of the quarks we don't understand. Physicists Ben Still reminds us that the quarks within protons are confined to a tiny space, so there is a large uncertainty to their momentum and energy.[18.] There is obviously more to the mass, energy, and freedom of movement of quarks yet to be discovered. If forces are proven to be due to the attraction and repulsion between fields, then it is likely gluons don't exist and that which is holding quarks together is either their respective fields or they are connected in Tor or Tryk Chains.

The three quarks in a nucleon normally appear with two of their spins aligned and one quark with its spin in the opposite direction. When that one odd quark flips over, resulting in all three quarks having their spins aligned it is called a *delta quark*, which otherwise has the appearance of any other nucleon. Physicist Timothy Paul Smith relates that the flipping of the third quark causes the proton's mass to jump from 938 MeV to 1232 MeV.[19.]

Dr. Still relates that the jump in mass/energy is due to a tension between the particles that is called *excited resonances.*[20.] He describes the phenomenon as trying to hold three, like-oriented, bar magnets in a close side by side arrangement. To hold them closely takes energy to overcome the natural tendency for the middle magnet to flip around, relieving the tension between the magnetic fields. When the third quark flips, it is unknown where the added mass comes from. Perhaps the center quark must acquire significant work energy to overcome the resultant tension and made the flip, and that added work energy raises the particle's angular momentum, giving it a stronger gyroscopic effect and hence more inertial mass.

Prediction: Once the viability of Penergy and spinning Tor particles is recognized, a clever astrophysicist will create the mathematics to demonstrate the relationship between the internal structure of matter, Tor spin, and the gyroscopic effect, providing a possible alternative definition of mass.

Resolution of
Cosmological Issues – Apndx.#1

☒ Explanation of Work Energy
☒ Foundation of Quantum Mechanics
☒ Magnetic Monopole Mystery
☒ Missing Anti-Matter Mystery
☒ Neutrino Oscillation
☒ Origin of Cosmic Web
☒ Origin of Dark Energy
☐ Origin of Dark Matter
☒ Origin of EM field
☒ Origin of EM Waves
☒ Origin of Fields
☒ Origin of First Stars
☒ Origin of Force
☐ Origin of Gravity
☒ Origin of Mass
☒ Particle Decay
☒ Particle Evolution Sequence
☒ Particle Tunneling
☒ Purpose of Entropy
☒ SMBH Creation
☐ Space & Time Relativity
☒ The Fine-Tuning Problem
☒ The Flatness Problem
☒ The Hierarchy Problem
☒ The Horizon Problem
☒ The Magnetic Field
☒ The Measurement Problem
☒ The Reality Problem
☒ Virtual Particles
☒ Wave-Particle Duality Resolution
☒ What Drives Cosmic Evolution
☒ What is Energy
☐ Why Gravity is so Weak
☒ Why Photons are Massless
☒ Why Three Generations

Our Penergy medium will likely contain odd particles and particle fragments that were stable enough to withstand the frenzy, but not flexible enough to become a part of something larger and more complex.

Appendix #2
The Origin of Dark Matter

In 1930 while observing a group of a few hundred galaxies bound together by their combined mass, Astronomer Fritz Zwicky noticed they were moving much faster than their combined mass could explain. He surmised the mass of galaxies contained a large amount of unseen additional matter, which he dubbed *dark matter*. Forty years later, Astronomer Vera Rubin, while studying the rotation speed of galaxies noticed that the outer stars in the galaxies were moving much faster than they should. She too surmised the influence came from an unseen form of dark matter. It has since been theorized the dark matter is a halo of mass around the galaxies, likely to be some undetected particle, though its type and source are a mystery. It is unknown whether dark matter really exists as there are other theories for the rotation-speed irregularities, but let's speculate on what that dark matter might be.

One of the consequences of the evolutionary process is the creation of a great deal of waste and debris from the relatively short lives of many evolutionary particle experiments. On earth, this debris is broken down in our oceans and soil. In space, unused particle matter is broken down by decay and collision, and that debris is left to drift with the cosmic tides.

Using Tors, Tor-Chains, and Tryk combinations, the evolutionary process could build all kinds and shapes of particles and anti-particles. Stable, long-term creations, however, are always limited by the hazards of the environmental frenzy and the competition for survival inherent in natural selection. As is often the case with natural selection, some of the stable creations that do not go on to develop further might continue to exist but remain developmentally stagnant. Accordingly, our Penergy medium will likely contain odd particles and particle fragments that were stable enough to withstand the frenzy, but not flexible enough to become a part of something larger and more complex.

Those odd Tors, Tryks, Chains, and particle fragments left behind in evolutionary development are unseen by us because only particles with electric charge can absorb/emit photons. Any particle remnant with an attractive or repulsive charge would float around, bumping into other fragments until it was annihilated or combined with the right fragment(s) to neutralize its charge and be unable to absorb/emit photons. Accordingly, it would be accurate to call these now neutral fragments collectively *dark matter.*

Physicists Paul Davies and John Gribbin expressed a similar idea in their book, *The Matter Myth.* On the origin of dark matter, they said, "Nobody is sure what the invisible stuff is, but the best bet is that it is an unseen residue of exotic subatomic particles left over from the big bang."[1] Some theorize dark matter to be some exotic new particle. It seems more reasonable for dark matter to be normal particle fragments, however, since it is observed that the density of dark matter has the same magnitude as the density of visible matter.[2]

These dark matter fragments would have floated around bumping into each other, gradually expelling their kinetic energy. No longer capable of combining or absorbing more energy, they were left to gather in clumps due to their own gravity. Scientists believe those clumps represent five-times more dark matter than the visible matter in the universe, and that its cumulative gravitational influence played a part in the shaping and distribution of the developing galaxies.

Much of the left-over particle debris would likely be in the form of very lengthy left and right-spinning Tor-Chains. Many were probably so long and complex that the likelihood of annihilation became remote. Consequently, as the cosmic turmoil around blackholes occasionally breaks up some of the dark matter, exposing those long chains, we would expect to see an unusual number of magnetic fields in their vicinity, which have been observed.

But these observations do not mean that dark matter exists in the volumes scientists currently believe. The invention of dark matter came about as a consequence of anomalies in rotational curves normally controlled by the gravitational field of the SMBH, so perhaps it is the SMBH gravity that is causing the anomalies. I don't think we currently know all there is to know about SMBH gravity. We will take a closer look at gravity in Appendix #3.

Resolution of
Cosmological Issues – Apndx.#2

- ☒ Explanation of Work Energy
- ☒ Foundation of Quantum Mechanics
- ☒ Magnetic Monopole Mystery
- ☒ Missing Anti-Matter Mystery
- ☒ Neutrino Oscillation
- ☒ Origin of Cosmic Web
- ☒ Origin of Dark Energy
- ☒ Origin of Dark Matter
- ☒ Origin of EM field
- ☒ Origin of EM Waves
- ☒ Origin of Fields
- ☒ Origin of First Stars
- ☒ Origin of Force
- ☐ Origin of Gravity
- ☒ Origin of Mass
- ☒ Particle Decay
- ☒ Particle Evolution Sequence
- ☒ Particle Tunneling
- ☒ Purpose of Entropy
- ☒ SMBH Creation
- ☐ Space & Time Relativity
- ☒ The Fine-Tuning Problem
- ☒ The Flatness Problem
- ☒ The Hierarchy Problem
- ☒ The Horizon Problem
- ☒ The Magnetic Field
- ☒ The Measurement Problem
- ☒ The Reality Problem
- ☒ Virtual Particles
- ☒ Wave-Particle Duality Resolution
- ☒ What Drives Cosmic Evolution
- ☒ What is Energy
- ☐ Why Gravity is so Weak
- ☒ Why Photons are Massless
- ☒ Why Three Generations

*The spinning and churning of the interior
and surrounding Penergy medium
contracts the surrounding medium
creating a gravitational field.*

Appendix #3
The Origin of Gravity

I love blackholes. They offer an endless variety of topics about which to speculate. I've long been interested in the idea of supermassive blackholes (SMBHs), but unimpressed with the idea of them forming from collapsing primal gases or the merger of multiple blackholes or galaxies. Those notions simply don't seem feasible given the billions of SMBHs in existence. SMBHs forming out of dense pure energy is a strange idea but it works out to be quite plausible. But that notion raises some questions, such as how does a pure-energy SMBH differ from the speck of pure energy from which our universe began?

Again, Penergy does not have mass until it is spun into a defined entity such as a particle, and therefore amorphous Penergy does not possess the capacity to exert gravity. This got me thinking about the presence of gravity without mass and the question of how a supergiant SMBH could ever be shrunk down to the size of a speck of pure energy. From that line of thinking I developed the notion of *dynamic gravity*, and from there I worked on the origin of gravity. There is no evidence for dynamic gravity, but it is an interesting speculation.

Einstein showed us that gravity causes a curvature of space in proportion to mass. I've often wondered, what is inherent in mass that causes the surrounding medium to bend? It doesn't make sense for the presence of a near-static mass alone to be capable of producing a gravitational field. Believing that a field is a tension in the intergalactic Penergy medium (IPM), I focused on what could be moving in the medium that could produce that tension. Once again, it turned out that the Tor particle was the only reasonable culprit. I developed the hypothesis that the gravitational field is much the same as a local field but on a regional basis. It is still Tor particle spin that is affecting the surrounding IPM and creating a field, but on a much bigger scale.

The Penergy medium is geometrically flat except where a gravitational influence causes a curvature.[1.] A mass sitting in space is surrounded by a gravitational field. Einstein gave us the mathematics to calculate how much curvature is taking place and to what degree that curvature will affect a change in the velocity of near-by masses. I visualize the surrounding medium as somehow being squeezed from all sides by a force directed toward the center of the mass. But, again, what is causing that spherical contraction of spacetime around the mass?

We know that massive, spinning objects such as blackholes curve space and time.[2.] It might be the *spin* of the object that is curving and contracting the Penergy medium, not simply the object's sheer mass. That reasoning may work for blackholes that are known to possess significant spin and are capable of dragging the surrounding space with it, but how would that work for particles that are only part of a massive object, especially one that may be spinning relatively slowly such as a planet? Where would its gravitational field come from? I will first speculate on the origin of the gravitational field of large masses, then blackholes, since the respective origins are different.

Hypothesis for the Gravitational Field of Large Masses

Recall that mass is simply an accumulation of spinning particles, and those particles are all comprised of Tor combinations. The only way any mass has contact with the Penergy medium is through the surface of those spinning Tors. Tors spin on their own axes and presumably at a very high rate, creating their own angular momentum (*Spin-Energy*) and field tension in the Penergy medium.

The fluid-like Penergy medium permeates the entire universe, filling every space in between SMBHs and particles. A solid mass is mostly empty space, but that space is actually filled with the Penergy medium and is present between every Tor, quark, nucleon, and atom. That Penergy medium permeating the interconnected space is spinning and churning created by the spin of all the Tors making up the mass. That spinning and churning occurs throughout the interior of the mass and carries over into the Penergy medium surrounding the mass.

Because the Tors are not soaring through space but are bound up in a mass, the Spin-Energy that would normally propel them through the Penergy medium is transformed into a strong, churning field energy. Instead of the Penergy pulling on the Tors to propel them, the Tors are pulling on the Penergy to contract it. The spinning and churning of the interior and surrounding Penergy medium contracts the surrounding medium, causing it to push inward, creating a gravitational field. As noted by Einstein, gravity does not pull, space pushes.[3.] The strength of the field is a function of the number of spinning Tors and the size of the body's mass – a value known as its compaction ratio. This is how mass is related to a gravitational field.

In conclusion, the hypothesis is that confined Tor-particle spin causes a strong tension in the Penergy medium both inside and outside the mass, pulling on the medium to curve and contract it, creating a negative pressure we recognize as a gravitational field. The strength of the gravitational field is in proportion to the object's mass – the number of Tors comprising the object.

In the case of neutron stars that have much less interconnected space between particle matter, there is still a very high concentration of spinning Tors. The overall size of the gravitational field is the same no matter the size of two objects having equal mass, but in the case of a neutron star with a very high compaction ratio, the gravity at its surface is very high.

This speculation is counter intuitive and there is no direct evidence for it, but there is observational evidence that suggests a connection between spin and gravity.

There are two types of mass: the measured resistance to a change in movement is called *inertial mass*, and the measure of how much gravitational force a mass exerts on another mass is called *gravitational mass*. These two types of mass are based on different concepts, but they always measure exactly the same. There is no obvious reason why these two masses should be the same until we realize that the nexus between the two is Tor spin and its inherent Spin-Energy.

As previously discussed, the source of inertial mass may be due to the gyroscopic effect created by the very fast spinning Tors. The strength of that gyroscopic effect is a function of Tor's inherent Spin-Energy, as defined in Section 3.4a. As described above, the source of the gravitational field from which the measure of mass is derived could also be due to the spinning of Tors and the strength of their field is also a function of their inherent Spin-Energy. Tor spin and its Spin-Energy appears to be the nexus between mass and gravity and the reason inertial mass and gravitational mass have the same value.

Further evidence associating spin and gravity is found in the fact that the strength of the gravitational force falls off at exactly the same rate as the electromagnetic force: inversely to the square of the distance between two masses or particles. The similarity at which the force diminishes between these two unrelated forces must be more than a coincidence. Two distinctly different forces that lose strength at the exact same rate strongly suggest they have some underlying feature in common.

One could argue that the commonality is that both forces are exchanging hypothetical massless bosons. But we have concluded that boson exchange to explain force is dubious, so the similarity in question must have another cause. Recall that the attraction/repulsion of the electric force is due to the dynamics of Tor spin. Now we are saying the nexus between mass and gravity could also be due to Tor spin. Tor spin being the source of both the electric field/force and gravitational field/force could be the reason both fields diminish in strength in the same proportion. Both are derived ultimately from the same source - Tor spin.

Prediction: Once the presence of Penergy, Tors, and personal fields is established, a physicist or astrophysicist will work out the mathematics to support, and the experiment to prove, that a gravitational field, like all other fields, is created by particle spin and its effect on the surrounding Penergy medium.

Hypothesis for the Gravitational Field of Blackholes

Blackhole mass is different than large-body mass, and the origin of their respective gravitational fields is different. That notion is not as strange as it may seem. As Physicist Brian Clegg points out, "It's only an assumption that gravitation behaves exactly the same on all scales."[4.]

Current theory ascribes blackhole mass to be related to a singularity at the center of the blackhole. I don't believe in physical singularities, and don't see how a singularity by itself could cause the contraction of the surrounding Penergy medium for such great distances. Again, for such a contraction of the Penergy medium to exist, something should be moving. This time, however, it couldn't be the Tor particle because inside the blackhole Tors no longer exist.

The strength of a blackhole's gravitational field is believed to be proportional to its mass, and its mass is calculated based on its gravitational effect on surrounding stars and gases. In other words, our calculation for a blackhole's mass is dependent on Einstein's equations to calculate its gravitational field. It is a presumption that blackhole mass is equivalent to large-body mass but apparently they are different, which could account for, at least in part, the discrepancies in galaxy rotational speeds.

As described at the beginning of Chapter Three, the measure of a blackhole's mass and gravitational field is proportional to the density of its Penergy content, which in turn has a relationship to its spin rate. This is quite a departure from current theory, but perhaps such a departure is necessary. As Astrophysicist Stuart Clark points out, "...general relativity is the ultimate gravitational theory. Yet it can't explain blackholes. If we believe that the universe is understandable through mathematics, the blackholes are telling us that there must be a deeper theory of gravity to be found."[5]

According to Einstein, mass couples to spacetime,[6] and as previously observed massive spinning objects such as blackholes are known to drag the surrounding spacetime Penergy medium. According to Physicist Raymond Phillips, frame dragging is a form of gravity – albeit a rather strange one.[7] Recall that a natural trait of the Penergy medium is that spin causes a tension and contraction. The spinning blackhole and frame dragging is setting up the conditions for the spin of the surrounding Penergy medium to cause the contraction of the medium and a gravitational field.

Recall from Chapter Three that matter inside a blackhole will likely go through transitional phases as it becomes denser. The blackhole crushes matter back into Penergy, the Penergy into denser and denser Penergy, and finally into its NIB (*Near Infinite But* not quite) state. The force of gravity as we know it can bring matter to the event horizon and even push it passed that point, but the actual crushing of matter back into Penergy may only occur inside the blackhole where Einsteins equations may no longer be valid.

The heart of the blackhole is a long distance from, and no longer influenced by, the gravitational curvature of the surrounding Penergy medium. Inside the blackhole there may be an influence forcing the Penergy toward the center, but that cause and the fate of the Penergy is now determined by the dynamics inherent in the internal Penergy condensing process. That Penergy condensing process is beyond the realm of our current idea of gravity described by General Relativity, so it will require further theory and perhaps new mathematics before it is understood.

Blackholes contain only Penergy and do not contain mass as we define it. True, the two are equivalent mathematically, and Penergy can spin itself into mass, but the *state* of the substance within a black-hole is amorphous, condensing Penergy, not physical mass. Amorphous Penergy is not equivalent to a concentrated mass and its presence will not create a gravitational field. The effects of gravity are not in play until the Penergy is spun into a particle and the spinning particle creates tension and contraction within the Penergy medium responsible for the gravitational field. Consequently, there is no gravity as we know it present within a SMBH. If amorphous Penergy alone created gravity, the early universe would not have been subject to expansion.

The radius to which a mass must be compressed down to form a blackhole is called the Schwarzschild radius, named after Karl Schwarzschild, who was first to find a solution to Einstein's equations describing a spherical blackhole. The bigger the mass of the blackhole, the larger the Schwarzschild radius. Under my hypothesis, this proportionality eventually changes.

As the Penergy condenses into a higher and higher density approaching its high-density NIB state, the SMBH shrinks causing it to spin faster. The faster it spins the greater its gravity. That is dynamic gravity. Dynamic gravity may not manifest itself until the blackhole is quite large and has processed a tremendous amount of matter and Penergy. The tell-tale sign of a NIB-state and dynamic gravity would be when the blackhole's spin rate goes up and its event horizon begins to shrink. Some SMBHs may be in the initial phase of that state already, as spin rates have been measured to be as much as ninety-five percent of light speed.[8.]

This phenomenon of added mass causing the circumference to shrink has been observed in neutron stars where the energy density is quite high. In contrasting the customary effects of gravity to neutron stars, Astronomer Luciano Rezzolla tells us that, '...the opposite is true, as the mass grows, the radius tends to decrease."[10.] Again, as Einstein pointed out, gravity doesn't pull, space pushes.

The spin-rate/circumference relationship is subject to the same dynamics an ice skater employs to spin faster by pulling in her arms, shrinking the radius of her mass, driving up the angular momentum and spin rate. In the blackhole's case, it is the faster spin rate that is causing the radius of the mass to shrink. Theoretically, if we fed all the matter and Penergy in the universe into a single, giant blackhole, its core would become denser, spin rate would continue to rise, its circumference would continue to shrink, and eventually it would return to the speck of NIB-density Penergy from which our universe began.

Stellar collapse goes through phases of compaction of its core and the shedding of its outer layers. The phase before becoming a blackhole is when it is a highly dense, fast spinning, neutron star. As the surrounding gravitational field compresses the star's final contents closer and closer together, it also causes the spin rate to climb. Eventually, the neutrons and all other stellar matter is crushed back into Penergy, and the gravitational field created by the spin of particles is transitioned into a gravitational field created by the spinning of the newly formed blackhole.

In conclusion, my hypothesis is that the gravitational field of a blackhole is created not directly by any form of mass within the blackhole, but by the contraction of the surrounding Penergy medium due to the spin of the blackhole.

If a blackhole's gravitational field is created by its spin rate, then the strength of that field should be less as one moves up and down from the equator. This, of course, is the reason a SMBH galaxy is shaped in a disk that is level with its equator, at least until a collision with another galaxy throws the disks in disarray. As one moves up and down from the equator the strength of the gravitational field diminishes, the number of stars on those planes grows fewer, creating a bulge of stars immediately surrounding the SMBH but few off-plane beyond. At the poles there is little gravitational influence, and they become the points where matter escapes in jets from the swirl of particles that surround the blackhole. There is no observational evidence for this hypothesis as yet, but I'm confident that careful observation of SMBHs will confirm that the gravitational field is strongest at a blackholes equator and weakest at its poles.

Prediction: Someday a sharp astrophysicist will figure out a way to measure the gravitational field surrounding a SMBH and discover that the strongest area of its gravitational field is at its equator, where it is spinning the fastest. This will confirm that it is not simply the mass of the SMBH that is creating its gravitational field, but the spinning of the SMBH.

Mond

Scientists have put forth theories that support stronger gravity and less reliance on dark matter to explain the unusual galaxy rotation mentioned in Appendix #1. Israeli Physicist Mordehai Milgrom proposed simple changes to Newton's laws that when applied to galaxies turned out to match their rotation speeds quite well. His theory is known as MOND for Modified Newtonian Dynamics. MOND theory is out there but is not yet well accepted. According to Physicist Brian Clegg, in the vast majority of galaxies, dark matter is less effective at matching the observed rotational curves than modified gravity.[10.] MOND, however, did show that the mass of a galaxy is related to its rotational velocity.[11.]

Mond and dark matter theories are feasible but may be rendered moot once we learn more about Penergy and the creation and dynamics of SMBHs. My hypothesis for blackhole gravity may affect galaxy rotation curves due to the gravitational field changing strength as one moves away from the equatorial plane. My hypothesis for the creation of the NIB state of highly condensed Penergy that drives up the SMBH spin rate would create a *Rankine vortex.* This ultra-high spin vortex was created by William John Macquorn Rankine who worked out the math that demonstrates such a vortex exhibits a rigid body rotation. The rigid body rotation applied to galaxy rotation could describe their irregular rotation speeds. These notions stem from my belief that there is much more to be learned from examining Penergy, the dynamics of its various densities, and its relationship to spin.

Conclusions

The magnetic, electric, and gravitational fields and their related forces, as well as centripetal and centrifugal mock-forces, are all derived from something spinning. Things that spin within the Penergy medium create a field. The basic difference between a particle field and gravitational field of a large mass is that the personal particle field is low in intensity and is produced by one particle which limits its influence to affecting other individual particles in its proximity. The large-mass gravitational field, however, is produced from the spin of a great many Tor particles giving it a higher intensity and a much larger area of influence depending on the number of particles involved and the density of the particle compaction. Both kinds of fields contract the surrounding Penergy medium establishing an area of influence.

Magnetic, electric, and gravitational fields all create an acceleration by virtue of their influence being inverse to the square of the distance between the source and its subject. The closer one gets to the source of the field, the stronger the field, accelerating one toward the source.

The attractive force of electric charge occurs directly between the fields of Left-Tors and Right-Tors, each directly sensing each other's field. However, the attractive force of gravity occurs between the Tor spin of large bodies and their cumulative effect of curving the surrounding Penergy medium, which *in turn* affects other particles. This indirect effect of gravity on other particles is likely the reason it is so much weaker than the other forces.

Some question whether gravity is a true force. It is. A field is a condition in the IPM that affects the velocity of nearby masses, and that effect is called a force. The gravitational field is a contraction in the IPM surrounding a mass or blackhole, and that contracted space can cause a change in the velocity of other masses, making that affect a force. Particles create fields and fields create forces. A gravitational field creates a gravitational force.

My hypothesis means that gravity is an emergent quality; it did not come about until particle spin was introduced into the universe.

Resolution of
Cosmological Issues – Apndx.#3

☒ Explanation of Work Energy
☒ Foundation of Quantum Mechanics
☒ Magnetic Monopole Mystery
☒ Missing Anti-Matter Mystery
☒ Neutrino Oscillation
☒ Origin of Cosmic Web
☒ Origin of Dark Energy
☒ Origin of Dark Matter
☒ Origin of EM field
☒ Origin of EM Waves
☒ Origin of Fields
☒ Origin of First Stars
☒ Origin of Force
☒ Origin of Gravity
☒ Origin of Mass
☒ Particle Decay
☒ Particle Evolution Sequence
☒ Particle Tunneling
☒ Purpose of Entropy
☒ SMBH Creation
☐ Space & Time Relativity
☒ The Fine-Tuning Problem
☒ The Flatness Problem
☒ The Hierarchy Problem
☒ The Horizon Problem
☒ The Magnetic Field
☒ The Measurement Problem
☒ The Reality Problem
☒ Virtual Particles
☒ Wave-Particle Duality Resolution
☒ What Drives Cosmic Evolution
☒ What is Energy
☒ Why Gravity is so Weak
☒ Why Photons are Massless
☒ Why Three Generations

Penergy density doesn't change "time" per se, only the movement of particles, and that rate of change in particle position (think atomic clocks) is how we measure time.

Appendix #4
Why Space & Time are Relative

The Penergy medium is a wonder to behold. Like the tough fabric it is, it can withstand the contractions, bending, and warping by massive blackholes, and at the same time act as stretched cellophane on which subtle vibrations can be transmitted allowing tiny particles to communicate. It can withstand the comparatively large gravitational waves created by crashing neutron stars, while at the same time holding a single particle's delicate field. It can hold the trillions of individual particles, their vibrational signals, all the photons of radiation and their related electromagnetic fields, and simultaneously allow the tiny signal from my friend's cell phone in Paris, via satellite, to find my cell phone in California, and relay her voice clearly and distinctly. The Penergy medium is truly dynamic with fantastic capabilities. It is also closely related to time and may be responsible for why space and time are relative to each other.

We can't see or feel space or time, but they remain very real to us. We derive their existence by sensations of objects and the movements of those objects around us. We once thought of space and time as distinct entities but have since learned from Albert Einstein that they are connected. Let's examine how that could be.

A Thought Experiment

Let's do a thought experiment. Imagine a capital T with a railroad track running across its top and a narrow conveyor belt running down its base. A mailman on a passing train wishes to toss a bag of mail onto the conveyor belt. The mail bag resting on the moving train has momentum, which will carry it forward as it falls toward the conveyor belt. Knowing the speed of the train and the drop distance to the conveyer belt and using Newton's laws of motion one can calculate when in advance to let go of the mail bag to hit the conveyer belt.

Now imagine the same setup with the capital T, but this time from the train we are going to emit a photon down the center of the conveyor belt. For this to occur, where should the train be when the photon is emitted? From all that I have read, the emission point on the train should be directly adjacent to the center of the conveyor belt. The photon will not travel forward in the direction of the train's motion as the mail bag did. When the photon comes into existence, it immediately steps out above the conveyor belt and shoots away in a *straight path* at light speed, and it will stay on that straight path until influenced by a gravitational field or some form of matter.

This makes sense because whereas the mail bag was *in existence* and subject to the laws of physics regarding the movement of it and the train, the photon as we know it *did not exist* in the same sense until it was emitted from the train.[1.] The photon was separate from the train the moment it came into existence, so the train had no influence on the photon's lateral movement.

The photon was only subject to the laws of physics regarding the Penergy medium, which is the only realm in which a photon exists. This goes along with the idea that the speed of light (*the photon*) is always measured the same, whether emitted from a source traveling in the direction of emission or away from the direction of emission. The speed and direction of the source of emission is immaterial to the photon.

Proof that photons travel only in straight lines irrespective of the movement of their source is found in experiments measuring the distance to the moon, where photons are shot to reflective mirrors left on the moon's surface. According to the experimenters, a light beam from earth that strikes the mirrors is reflected back on the exact same path that the beam took to reach the mirrors.[2.]

If the light beam shot from a spinning earth is reflected on the same path from the non-spinning moon, then obviously the spin of the earth had no effect on the path of the beam. It is a straight shot in both directions. As physicist Brian Clegg says, "...light always goes in a straight line. Period. There is no arguing with this."[3.]

Another Thought Experiment

The conclusion drawn from the above thought experiment is that a light beam may bend due to gravity, but it will not bend due to the movement of the source of the beam. If this is true, it brings into question an often-used argument for time dilation -- the light-beam-bouncing-within-a-moving-spaceship scenario.

The argument is that two mirrors perfectly parallel to each other are set up on the floor and ceiling of a spaceship with a light gun mounted at the upper mirror aimed at the lower mirror. Once the spaceship is under way at high speed the light gun fires light (*photons*) at the bottom mirror. The light hits the bottom mirror and is reflected up, hitting the top mirror and reflected down, etc.

From inside the spacecraft the photons are simply seen bouncing up and down between the mirrors. For an observer outside the spaceship, however, she sees the same light moving in the direction of the spaceship making a series of W's between the mirrors. Given the light is moving in a W, a distance much greater than the direct distance between the mirrors, the photons must be traveling a much longer distance from mirror to mirror. Because light only travels at one speed, the only way for it to travel the longer distance is for *time to slow* within the fast-moving spaceship.[4.]

At first impression one may wonder how the photons traveling a lateral distance could be said to be traveling a greater distance necessitating more time than it takes to bounce them up and down between the mirrors. The time it takes for a bullet dropped from five feet to hit the ground is the same for a bullet fired from a five-foot-high level rifle, even though the bullet may travel a thousand yards laterally before striking the ground. Lateral movement does not add any time to falling bullets, why should it for bouncing photons? I must be missing something here.

That argument aside, let's continue with our thought experiment and say we make the bottom mirror very tiny, perhaps only a few atoms wide. It is still parallel to the top mirror and still capable of reflecting the light gun photons. We again get the spaceship up to speed and fire the light gun. The light travels toward the bottom mirror, but this time it does not reflect up to the top mirror. It missed hitting the tiny bottom mirror because the movement of the spaceship moved the bottom mirror out of the way before the light reached it.

In the absence of a gravitational field that might bend the path of the photons, they will move in a straight line downward and not be influenced in any way by the movement of the spacecraft. The light when emitted from the light gun was set on a *straight path*, and but for the high-speed spaceship moving the bottom mirror out of the way, the light would have struck the mirror and been reflected. Having missed the bottom mirror, the light does not move in a W or travel a greater distance by being emitted from inside a moving spaceship. For it to be reflected, the light would have to move in an arc back toward the bottom mirror. If that were the case, light would bend in an arc every time it was emitted from the spinning earth, but as discussed above, apparently that is not the case.

Relativity of Space and Time

The reflected photons scenario may not serve as proof for time dilation, but time dilation is nonetheless real. The scenario may, however, provide a means to demonstrate which of two passing spaceships is actually moving. It will be the one in which the photon gun misses its target. I am confident that some day an experiment will prove that point.

According to Einstein, the gravitational field of a large mass will cause time to slow. Time is relative to gravity as well as space due to the equivalency between a gravitational field and acceleration. Perhaps time is not relative to space directly, but relative to something equivalent to space that includes gravity. Let's explore what that could be.

Space is the void between stars and galaxies. We call it the vacuum of space, but we now know there is no such vacuum. Space is permeated with dark energy, which we have identified as Penergy. So, when we are referring to *space* or *spacetime fabric*, we are really referring to the Penergy medium.

For Penergy and time to be connected and relative to each other, it would mean that time is somehow a dynamic of Penergy, perhaps a function of its *density*. Time is relative to something changing, and it may well be that Penergy density affects how quickly change takes place. That may sound crazy, but let's examine the notion and see where it takes us.

If time were a function of Penergy density, then a change in density would mean a change in the passage of time. As we have seen,

Penergy was once very dense, but it has been thinning since the universe began. Its density now only changes to a higher density when it is condensed or compressed. Let's examine how those conditions could come about and whether they are related to a change in time. Einstein identified two situations causing time to change: *when near a mass with a strong gravitational field*, and *when flying through space at a very high rate of speed*.

Time Dilation Due to Gravitational Condensing

Einstein's general theory of relativity taught us that the gravitational field of a mass affects both space and time, and that the greater the gravitational field the more space is curved and the slower time passes.

One of the proofs of Einstein's theory is that the GPS satellites orbiting the earth must adjust for the time-passage differential between the earth's surface and their orbital height above the surface, otherwise the system would be thrown off significantly within a short time. This is due to the clocks in the satellites experiencing a lesser gravitational-field influence than the clocks on the surface of the earth. Time is running slower on the surface than it is in orbiting satellites twenty thousand kilometers above the earth's surface.

As we have established, a body with mass curves the geometry of the surrounding spacetime, creating what appears as an attractive influence. More accurately, it is the contraction of the surrounding Penergy medium that presses against the surface of the mass. The contraction raises the density of the surrounding IPM in proportion to the mass's gravitational field. That higher density of Penergy medium could be what is indirectly causing time to slow, not simply the gravitational field produced by the mass.

In the case of a blackhole, the strength of the gravitational field is enormous, having a significant influence on time passage. Theoretically the crew on a spaceship heading into a blackhole would pass through the event horizon without noticing any time change. A person watching the spaceship from a distance, however, would see time in reference to the craft gradually slowing and seemingly come to a stop as the spacecraft reaches the event horizon.[5.] That notion has obviously not been validated, but the point is that the stronger the gravitational field, the slower time passes, even at extremes.

We have witnessed particle mass being drawn into and consumed by a blackhole. The Penergy medium, however, is drawn close but its high rate of movement around the blackhole gives it angular momentum, which pushes the Penergy medium away at the event horizon. While the blackhole is rapidly spinning, the Penergy medium is pulled in tightly to the blackhole but held at bay. Obviously, if fast spinning blackholes eat Penergy like they do objects with mass, the first blackholes would have consumed all the remaining Penergy in the universe and we would not be here.

A likely hypothesis is that fast-spinning blackholes drag the surrounding Penergy medium but consume little of it. The surrounding Penergy builds in density as it is pulled close, causing time to slow as seen by the observer watching the rocket approaching the event horizon.

The much higher Penergy density at the event horizon likely also causes particle pairs to be created consistent with the strength of the density. Those particle pairs include the first and second generation of particles – the heavier quarks and leptons. Some of those particle pairs would be drawn into the blackhole, and some would escape, being slung out into space. This could be why the earth is constantly bombarded by cosmic rays – high energy particles and anti-particles from space having an indeterminant origin. I visualize that someday when we have the technology to travel very quickly through space, our spaceship will compress the Penergy at its leading edge to the extent the raised Penergy density will create particles. Given we possess the technology allowing the spacecraft to absorb those particles, we will be producing our own source of raw material from which to fuel our existence in space.

Time Dilation Due to High-Speed Compression

The other circumstance in which time is dilated according to Einstein is when a mass such as a rocket is traveling rapidly through spacetime. The closer the rocket is traveling to the speed of light, the slower time passes for those within the rocket, the rocket would grow in mass, and it would shrink in the direction of travel.[6]

The high-speed mass moving through the spacetime Penergy medium would cause the Penergy around it to compress. The compression would raise the surrounding Penergy density in proportion to both the mass's speed and mass-density. The change in the spacecraft's mass-density would be enormous. A mere particle traveling at .9999% light speed would have a density 10,000 times greater than its rest mass.[7] Consequently, the combined speed and greater mass-density would compress the surrounding Penergy, raising its density, and slowing time proportionally for those inside the rocket.

The same effect may even apply to something as small as a particle. A particle speeding through our ubiquitous Penergy medium would also compress the medium, raising the density around it and slowing time for the particle. The slowing of time means that the spin rate of a photon would slow down proportionally, lengthening the wavelength produced by its field. That would be reflected in a redshift in proportion to the time the photon would be gradually losing spin rate as it soared across our universe. It's only speculation, but it's an interesting idea.

Einstein pulled together space and time into one concept. We can see now why that works; space is made up of Penergy, and Penergy density affects the time passage, possibly including the rate of particle movements. Physicist Wouter Schmitz implies this relationship when in discussing the movement of mass through the vacuum he says, "So, the elasticity of the vacuum governs how fast we can move."[8.] I am simply saying that it is the density of the vacuum (Penergy) that governs how fast we can move. Penergy density doesn't change "time" per se, only the movement of particles, and that rate of change in particle position (*think atomic clocks*) is how we measure time.

There is no hard evidence for either of these time dilation theories being related to Penergy density. Curiously, however, in a July 2023 article in *Live Science*, Ben Turner reports that scientists have found new evidence for time having passed much slower in the early universe. This arose out of a twenty-year study of twelve-billion-year-old quasars and the analysis of the different wavelengths they emitted. The findings are said to have finally proven a prediction made by Albert Einstein.[9.] The Tor Model posits the Penergy density being higher in the early universe, even after SMBHs were created, so there may be a correlation with time passing slower. Notwithstanding this scant evidence, the relationship between Penergy density and time dilation represents a possible nexus between time, spin, gravity, and space. If proven valid someday, it would mean that the Penergy medium truly is *spacetime*.

Resolution of
Cosmological Issues – Apndx.#4

- ☒ Explanation of Work Energy
- ☒ Foundation of Quantum Mechanics
- ☒ Magnetic Monopole Mystery
- ☒ Missing Anti-Matter Mystery
- ☒ Neutrino Oscillation
- ☒ Origin of Cosmic Web
- ☒ Origin of Dark Energy
- ☒ Origin of Dark Matter
- ☒ Origin of EM field
- ☒ Origin of EM Waves
- ☒ Origin of Fields
- ☒ Origin of First Stars
- ☒ Origin of Force
- ☒ Origin of Gravity
- ☒ Origin of Mass
- ☒ Particle Decay
- ☒ Particle Evolution Sequence
- ☒ Particle Tunneling
- ☒ Purpose of Entropy
- ☒ SMBH Creation
- ☒ Space & Time Relativity
- ☒ The Fine-Tuning Problem
- ☒ The Flatness Problem
- ☒ The Hierarchy Problem
- ☒ The Horizon Problem
- ☒ The Magnetic Field
- ☒ The Measurement Problem
- ☒ The Reality Problem
- ☒ Virtual Particles
- ☒ Wave-Particle Duality Resolution
- ☒ What Drives Cosmic Evolution
- ☒ What is Energy
- ☒ Why Gravity is so Weak
- ☒ Why Photons are Massless
- ☒ Why Three Generations

Notes & References

Preface

1. Devereux, Carolyn, *Cosmological Clues*, 2021, p.35
2. Kaku, Michio, *The God Equation - The Quest for a Theory of Everything*, 2021

Introduction

1. Clark, Stuart, *The Unknown Universe*, 2016, p.287
2. Cliff, Harry, *Space Oddities*, 2024, p.26
3. Clegg, Brian, *Gravity*, 2012, p.186
4. Cliff, Harry, *Space Oddities*, 2024, p.263
5. Carroll, Sean, *Biggest Ideas-Space, Time and Motion, 2022,* p.175
6. Ekeberg, Bjorn, *Metaphysical Experiments*, 2019, p.151

Chapter One
1.0 Fields & the Double-Slit Experiment

1.1 The Origin of Particle Fields

1. Carroll, Sean, *The Big Picture,* 2017, p.173
2. Munowitz, Michael, *Knowing,* 2005, p.38
3. Musser, George, *Spooky Action at a Distance,* 2015, p.134
4. Munowitz, Michael, *Knowing,* 2005, p.322
5. Carroll, Sean, *The Big Picture,* 2017, p.176
6. James, Tim, *Fundamental,* 2020, p.50
7. Nomura & Poirier & Terning, *Quantum Physics, Mini Black Holes, and the Multiverse,* 2018, p.16
8. Geach, James, *Five Photons,* 2018, p.73

1.2 The Double-Slit Experiment
9. Ford, Kenneth, *The Quantum World,* 2005, p.196
10. Smolin, Lee, *Einstein's Unfinished Revolution,* 2019, p.100
11. Becker, Adam, *What is Real,* 2019, p.89
12. Ball, Philip, *Beyond Weird,* 2020, p.110

Chapter Two
2.0 Choosing the Initial Conditions
1. Cliff, Harry, *How to Make an Apple Pie from Scratch*, 2021, p.302

2.1 Initial Conditions
2. Devereux. Carolyn, *Cosmological Clues*, 2021, p.73
3. Hawking, Stephen, *A Brief History of Time*, 1988, p.50
4. Kaku, Michio, *The God Equation*, 2021, p.57
5. Impey, Chris, *Einstein's Monsters*, 2019, p.14
6. Sutter, Paul, *Your Place in the Universe*, 2018, p.31
7. Devereux, Carolyn, *Cosmological Clues*, 2021, p.73

2.2 What is Energy
8. Feynman, Richard P., *Six Easy Pieces,* 1996, p.71

2.3 An Overview of Penergy
9. Kaku, Michio, *The God Particle*, 2021, p.41

Chapter Three
3.0 The Origin of Supermassive Blackholes
3.1 Blackholes
1. Hawking, Stephen, *Blackholes and Baby Universes*, p.75
2. Rezzolla, Luciano, *The Irresistible Attraction to Gravity,* 2023, p.107
3. Natarajan, Priyamvada, *Mapping the Heavens*, 2016, p.92

3.2 Arguments for and Against SMBH Creation by Mergers
4. Impey, Chris, *Einstein's Monsters*, 2019, p.146
5. Impey, Chris, *Einstein's Monsters*, 2019, p.103
6. Impey, Chris, *How it Began*, 2012, p.165
7. Impey, Chris, *How it Began*, 2012, p.152
8. Jones, Mark, *An Introduction to Galaxies and Cosmology*, 2015, p.92

3.3 Alternative Theory for SMBH Creation
9. Barrow & Silk, *The Left Hand of Creation,* 1983, p.140 & 152

3.5 Evidence for SMBH Creation & the Cascade Effect

10. Barnes, *The Cosmic Revolutionary's Handbook*, 2020, p.240
11. Impey, Chris, *How it Began*, 2012, p.196
12. Sutter, Paul, *Your Place in the Universe,* 2018, p.162
13. Sutter, Paul, *Your Place in the Universe,* 2018, p.165
14. Gribbin, John, *In the Beginning*, 1993, p.212-213
15. Impey, Chris, *Einstein's Monsters*, 2019, p.143
16. Tucker, *Chandra's Cosmos*, 2017, p.61
17. Devereux, *Cosmological Clues*, 2021, p.108

Chapter Four
4.0 The Intergalactic Medium

4.2 The Mystery of the Missing Anti-Matter

1. Tyson, Neil deGrasse, *Origins,* 2014, p.26
2. Cham & Whiteson, *We Have No Idea*, 2018, p.212
3. Kaku, Michio, *Physics of the Impossible*, 2008, p.247
4. Cham & Whiteson*, We Have No Idea,* 2018, p.212
5. Close, Frank, *Antimatter,* 2018, p.84
6. Schumm, Bruce, *Deep Down Things*, 2004, p.124
7. Butterworth, Jon, *Atom Land*, 2019, p.120
8. Schumm, Bruce, *Deep Down Things*, 2004, p.98
9. Smolin, Lee, *The Trouble with Physics*, 2007, p.73

4.3 The Intergalactic Medium

10. Sutter, Paul, *Your Place in the Universe*, 2018, p.86
11. Livio, Mario, *The Accelerating Universe*, 2000, p.126
12. Randall, Lisa, *Knocking on Heaven's Door,* 2012, p.99
13. Mee, Nicholas, *Gravity*, 2022, p.279
14. Cham & Whiteson *, We Have No Idea*, 2018, p.90
15. Cham & Whiteson *, We Have No Idea*, 2018, p.103
16. Schmitz, Wouter, *Particles, Fields and Forces,* 2019, p.17
17. Mersini-Houghton, *Before the Big Bang*, 2022, p.70
18. Davies, Paul, *What's Eating the Universe*, 2021, p.25

Chapter Five
5.0 The Theoretical Basis for Force

1. Beckman, Milo, *Math Without Numbers*, 2021, p.191
2. Kaku, Michio, *The God Equation*, 2021, p.101

5.1 Force According to the Standard Model

3. Hawking, Stephen, *A Brief History of Time,* 1988, p.68-69

4. Butterworth, Jon, *Atom Land*, 2019, p.217
5. Hawking, Stephen, *A Brief History of Time*, 1988, p.157
6. Kaku, Michio, *The God Equation*, 2021, p.104
7. Ford, Kenneth, *The Quantum World,* 2005, p.86-87
8. Musser, George, *Spooky Action at a Distance,* 2015, p.73
9. Krauss, Lawrence, *The Greatest Story Ever Told,* 2018, p.104
10. James, Tim, *Fundamental*, 2020, p.180-181
11. Kaku, Michio*, The God Equation*, 2021, p.138

5.2 Force According to the Tor Model

12. Wilczek, Frank, *A Beautiful Question*, 2015, p.260
13. Krauss, Lawrence, *The Greatest Story Ever Told,* 2018, p.253
14. Munowitz, Michael*, Knowing*, 2005, p.32
15. Davies, Paul, *What's Eating the Universe, 2021 p.59*

Chapter Six
6.0 The Early Cosmic Environment

6.2 Evolution Progresses in Levels

1. Davies, Paul, *The Cosmic Blueprint*, 1989, p.5

6.4 Emergent Qualities

2. Alexander, Stephon, *Fear of a Black Universe*, 2021, p.65

6.5 Cosmic Homogeneity

3. Devereux, Carolyn, *Cosmological Clues*, 2021, p.116
4. Hawking & Mlodinow, *The Grand Design,* 2010, p.129
5. Devereux, Carolyn, *Cosmological Clues*, 2021, p.74
6. Greenstein, George*, Symbiotic Universe,* 1988, p.164
7. Guth, Alan H., The Inflationary Universe, 1997, p.238
8. Rothman, Tony, *A Little Book about the Big Bang,* 2022, p.161
9. Carroll, Sean, *The Big Picture*, 2017, p.308

Chapter Seven
7.0 Configurations: Photon & Neutrino

7.1 Arguments for Composite Particles

1. Schmitz, Wouter, *Particles, Fields, and Forces,* 2019, p.155
2. Schumm, Bruce, *Deep Down Things*, 2004, P.187
3. Fritzsch, Harald, *The Fundamental Constants*, 2005, p.27
4. Fritzsch, Harald, *The Fundamental Constants*, 2005, p.116
5. Barrow & Silk, *The Left Hand of Creation*, 1993, p.142
6. Lindley, David, *The Dream Universe*, 2020, P.136

7. Lincoln, Don, *Einstein's Unfinished Dream*, 2023, p.239
8. Butterworth, Jon, *Atom Land*, 2019, p.275
9. Munowitz, Michael, *Knowing*, 2005, p.34
10. Hawking, Stephen, *A Brief History of Time*, 1988, p.167
11. Cham & Whiteson, *We Have No Idea*, 2018, p.50
12. Impey, Chris, *How it Began*, 2012, p.298
13. Cham & Whiteson, *We Have No Idea*, 2018, p.54

7.2 Particle Configurations
14. Kisak, Paul, *The Photon*, p.5-6

Chapter Eight
8.0 Configs: Quark, Electron, Nucleus & Atom
1. Butterworth, Jon, *Atom Land*, 2019, p.137

8.2 Quark Configurations
2. Still, Ben, *Particle Physics Brick by Brick*, 2018, p.121

8.3 Electron Configuration
3. Wilczek, Frank, *Fundamentals*, 2021, p.89
4. Alexander, Stephon, *Fear of a Black Universe*, 2021, p.70

8.4 The Nucleus and Atom
5. Perlov, Mario, *Cosmology for the Curious*, 2017, p.189-190
6. Baggott, Jim, *Origins*, 2015, p.66

8.5 Observation Evidence for BBT
7. Devereux, Carolyn, *Cosmological Clues*, 2021, p.3

Chapter Nine
9.0 The Reality Puzzle
1. Smolin, *The Trouble with Physics*, 2007, p.7

9.1 Finding Reality
2. Hertog, Thomas, *On the Origin of Time*, 2023. P.176
3. Wilczek, Frank, *A Beautiful Question*, 2015, p.9
4. Kaku, Michio, *Visions*, 1997, p.109

9.2 Quantum Theory
5. Nomura, Poirier & Terning, *Quantum Physics, Mini Black Holes, and the Multiverse*, 2018, p.21
6. Krauss, Lawrence, *The Greatest Story Ever Told*, 2018, p.91
7. Smolin, Lee, *The Trouble with Physics*, 2007, p.7

8. Nomura, Poirier & Terning, *Quantum Physics, Mini Black Holes, and the Multiverse*, 2018, p.21
9. Nomura, Poirier & Terning, *Quantum Physics, Mini Black Holes, and the Multiverse*, 2018, p.21
10. Hertog, Thomas, *On the Origin of Time*, 2023, p.192
11. Hertog, Thomas, *On the Origin of Time*, 2023, p.194
12. Hertog, Thomas, *On the Origin of Time*, 2023, p.194
13. Becker, Adam, *What is Real*, 2019, p.150
14. Nomura, Poirier & Terning, *Quantum Physics, Mini Black Holes, and the Multiverse*, 2018, p.63
15. Clegg, Brian, *The God Effect*, 2006, p.34
16. Becker, Adam, *What is Real*, 2019, p.150-151
17. Smolin, Lee, *Einstein's Unfinished Revolution*, 2019, p.128
18. Nomura, Poirier & Terning, *Quantum Physics, Mini Black Holes, and the Multiverse*, 2018, p.64
19. Nomura, Poirier & Terning, *Quantum Physics, Mini Black Holes, and the Multiverse*, 2018, p.38
20. Carroll, Sean, *Something Deeply Hidden*, 2020, p.20
21. Ball, Philip, *Beyond Weird*, 2020, p.114
22. Becker, Adam, *What is Real*, 2019, p.17

9.3 The Reality of Quantum Theory

23. Carroll, Sean, *Something Deeply Hidden,* 2020, p.18
24. Nomura, Poirier & Terning, *Quantum Physics, Mini Black Holes, and the Multiverse,* 2018, p.63
25. Kaku, Michio, *Visions*, 1997, p.351
26. Becker, Adam, *What is Real,* 2019, p.97
27. Rovelli, Luciano, *Reality is Not What it Seems,* 2017, p.178
28. Becker, Adam, *What is Real,* 2019, p.5
29. Carroll, Sean, *Something Deeply Hidden,* 2020, p.38-40
30. James, Tim, *Fundamental*, 2020, p.47
31. Carroll, Sean, *Something Deeply Hidden,* 2020, p.131-132
32. Carroll, Sean, *Something Deeply Hidden,* 2020, p.35
33. Hertog, Thomas, *On the Origin of Time,* 2023, p.177
34. Smolin, Lee, *Einstein's Unfinished Revolution,* 2019, p.29
35. Ford, Kenneth, *The Quantum World,* 2005, p.128
36. Nomura, Poirier & Terning, *Quantum Physics, Mini Black Holes, and the Multiverse,* 2018, p.3
37. Jones. Mark H., Robert J.A. Lambourne and Stephen Serjeant, *Intro. to Galaxies and Cosmology*, 2015, p.253

Chapter Ten
10.0 The Big Picture

10.1 In Search of the Big Picture

1. Randall, Lisa, *Knocking on Heaven's Door*, 2012, p.253

10.2 Cosmic Perspectives

2. Perlov, Delia, and Alex Vilenkin, *Cosmology for the Curious, 2017,* p.301-302.
3. Hawking, Stephen & Leonard Mlodinow, *The grand Design,* 2010, p.162.
4. Baggott, Jim, *Origins*, 2015, p.4.
5. Cliff, Harry, *How to Make an Apple Pie from Scratch*, 2021, p.339
6. Krauss, Lawrence M., *A Universe from Nothing*, 2012, p.89.
7. Hossenfelder, Sabine, *Existential Physics,* 2022, p.20

10.3 Cosmic Reality

8. Rezzolla, Luciano, *The Irresistible Attraction of Gravity,* 2023, p.280.·
9. Lincoln, Don, *Einstein's Unfinished Dream*, 2023, p.280.

Appendix

#1 Origin of Particle Mass

1. Baggott, Jim, *Quantum Reality,* 2020 , p.75
2. Wilczek, Frank, *Fundamentals*, 2021, p.57
3. Dine, Michael, *This Way to the Universe*, 2022, p.138
4. Krauss, Lawrence, *The Greatest Story Ever Told,* 2018, p.271
5. Kaku, Michio, *The God Equation*, 2021, p.89
6. Hossenfelder, Sabine, *Lost in Math,* 2020, p.37-38
7. Still, Ben, *Particle Physics Brick by Brick*, 2018, p.163
8. Krauss, Lawrence, *The Greatest Story Ever Told,* 2018, p.255-256
9. Krauss, Lawrence, *The Greatest Story Ever Told,* 2018, p.256
10. Fritzsch, Harald, *The Fundamental Constants*, 2005, p.113
11. Lindley, David, *The Dream Universe*, 2020, p.151
12. Lincoln, Don, *Einstein's Unfinished Dream*, 2023, p.42
13. Krauss, Lawrence, *The Edge of Knowledge*, 2023, p.88
14. Impey, Chris, *Einstein's Monsters,* 2019, p.174
15. Wilczek, Frank, *Fundamental*, 2021, p.74
16 Lange, Marc, *An Introduction to The Philosophy of Physics*, 2002, p.169
17. Wilczek, Frank, *Fundamentals*, 2021, p.74
18. Still, Ben, *Particle Physics Brick by Brick*, 2018, p.117

19. Smith, Timothy Paul, *Hidden Worlds*, 2003, p.63
20. Still, Ben, *Particle Physics Brick by Brick,* 2017, p.105

#2 Origin of Dark Matter
1. Davies & Gribbin, *The Matter Myth*, 1992, p.176
2. Alexander, Stephon, *Fear of a Black Universe*, 2021, p.113

#3 Origin of Gravity
1. Mee, Nicholas, *Gravity,* 2022, p.20
2. Impey, Chris, *Einstein's Monsters*, 2019, p.174
3. Kaku, Michio, *The God Particle,* 2021, p.43
4. Clegg, Brian, *Dark Matter & Dark Energy*, 2019, p.64
5. Clark, Stuart, *The Unknown Universe*, 2016, p.162
6. Impey, Chris, *Einstein's Monsters,* 2019, p.174
7. Philips, Raymond G., *Truth and absurdity in Modern Physics*, 2021, p.195
8. Impey, Chris, *Einstein's Monsters,* 2019, p.176
9. Rezzolla, *The Irresistible Attraction to Gravity,* 2023, p.98
10. Clegg, Brian, *Dark Matter & Dark Energy,* 2019, p.137
11. Devereux, Carolyn, *Cosmological Clues*, 2021, p.91

#4 Why Space & Time are Relative
1. Krauss, Lawrence, *The Greatest Story Ever Told*, 2018, p.81.
2. Cowen, Ron, *Gravity's Century*, 2019, p.93
3. Clegg, Brian, *Gravity*, 2016, p.128
4. Munowitz, Michael, *Knowing*, 2005, p.106
5. Clegg, Brian, *Gravity*, 2016, p.174
6. Close, Frank, *Antimatter*, 2018, p.28.
7. Gribbin, John, *In the Beginning*, 1993, p.223
8. Schmitz, Wouter, *Particles, Fields & Forces*, p.45
9. Turner, Ben, *Time Moved Five Times Slower*, Live Science Magazine, July 2023

Bibliography

Alexander, Stephon, *Fear of a Black Universe – An Outsiders Guide to the Future of Physics*, Basic Books, New York, 2021.

Baggott, Jim, *Origins - The Scientific Story of Creation*, Oxford University Press, 2015.

Baggott, Jim, *Quantum Reality – The Quest for the Real Meaning of Quantum Mechanics – A Game of Theories*, Oxford University Press, New York, 2020.

Ball, Philip, *Beyond Weird – Why Everything You Thought About Quantum Physics is Different*, University of Chicago Press, Chicago, IL, 2018.

Barnes, Luke A. & Geraint F. Lewis, *The Cosmic Revolutionary's Handbook – Or: How to Beat the Big Bang*, Cambridge University, 2020.

Barrow, John D. and Joseph Silk, *The Left Hand of Creation -- The Origin and Evolution of the Expanding Universe*, Oxford University Press, New York, 1993.

Becker, Adam, *What is Real -- The Unfinished Quest for the Meaning of Quantum Physics*, Basics Books, New York, 2019.

Beckman, Milo, *Math Without Numbers*, Dutton/Penguin Random House LLC, New York, 2021.

Butterworth, Jon, *Atom Land – A Guided Tour Through the Strange (and Impossibly Small) World of Particle Physics,* First Paperback Edition, The Experiment, New York, 2019.

Carroll, Sean, *The Biggest Ideas in the Universe – Quanta and Fields,* Dutton – Penguin Random House, New York, 2024.

Carroll, Sean, *The Biggest Ideas in the Universe – Space, Time, and Motion,* Dutton – Penguin Random House, New York, 2022.

Carroll, Sean, *The Big Picture – The Origins of Life, Meaning, and the Universe Itself,* Dutton, an imprint of Penguin Random House, LLC, New York, Paperback Edition, 2017.

Carroll, Sean, *Something Deeply Hidden – Quantum Worlds and the Emergence of Spacetime*, Dutton, 2019. Hard cover edition. A good read for understanding Quantum Mechanics, the arguments for a Wave Function, Spacetime, Quantum Field Theory, Entanglement, and Multi-verses.

Carroll, Sean, *Something Deeply Hidden – Quantum Worlds and the Emergence of Spacetime*, Dutton, paperback edition, 2020.

Chaisson, Eric J., *Cosmic Evolution – The Rise of Complexity in Nature*, Harvard University Press, Cambridge, MA, 2001.

Cham, Jorge & Daniel Whiteson, *We Have No Idea – A Guide to the Unknown Universe,* Riverhead Books, New York, Paperback Edition, 2018.

Clark Stuart, *The Unknown Universe – A New Exploration of Time, Space, and Modern Cosmology*, Pegasus Books, New York, 2016.

Clegg, Brian, *Gravity*, Duckworth Overlook, paperback edition, 2016.

Clegg, Brian, *The God Effect – Quantum Entanglement, Science's Strangest Phenomenon*, St. Martin's Press, New York, 2006.

Cliff, Harry, *How to Make an Apple Pie from Scratch*, Double Day, New York, 2021.

Cliff, Harry, *Space Oddities – The Mysterious Anomalies Challenging Our Understanding of the Universe,* Double Day, New York, 2024.

Close, Frank, *Antimatter*, Oxford University Press, United Kingdom, second edition paperback, 2018.

Close, Frank, *The Infinity Puzzle – Quantum Field Theory and the Hunt for an Orderly Universe,* Basic Books, New York, 2011.

Davies, Paul, The Cosmic Blueprint – New Discoveries in Nature's Creative Ability to Order the Universe, Simon & Schuster, New York, Touchstone Edition, 1989.

Davies, Paul, *What's Eating the Universe – and Other Cosmic Questions,* University of Chicago Press, 2021.

Davies, Paul and John Gribbin, *The Matter Myth - Dramatic Discoveries that Challenge our Understanding of Physical Reality*, Simon & Schuster, New York, 1992.

Devereux, Carolyn, *Cosmological Clues – Evidence for the Big Bang, Dark Matter, and Dark Energy*, CRC Press, Taylor & Francis Group, Baca Raton, Florida, 2021.

Dine, Michael, *This Way to the Universe*, Dutton/Penguin Random House, LLC, 2022.

Enos, Roland, *The Science of Spin,* Scribner/Simon & Schuster, New York, NY, 2023.

Feynman, Richard P., *Six Easy Pieces,* Addison Wesley Publishing Co., Reading, Mass.. 1995.

Ford, Kenneth W., *The Quantum World - Quantum Physics for Everyone*, Cambridge, Mass.: Harvard University Press, Copyright © 2004, 2005 by the President and Fellows of Harvard College.

Fritzsch, Harald, *The Fundamental Constants – A Mystery of Physics*, World Scientific, Singapore, 2005.

Geach, James*, Five Photons – Remarkable Journeys of Light Across Space and Time*, Reaktion Books, London, 2018.

Gott, Richard, *Cosmic Web – Mysterious Architecture of the Universe*, Princeton University Press, Princeton, NJ Paperback Edition, 2018.

Gould, Roy R., *Universe in Creation – A New Understanding of the Big Bang and the Emergence Life*, Harvard University Press, London, 2018.

Greene, Brian, *Until the End of Time – Mind, Matter, and our Search for Meaning in an Evolving Universe*, Alfred A Knopf, New York, 2020.

Gribbin, John, *In the Beginning – After COBE and Before the Big Bang*, Little, Brown and Co., New York, 1993.

Guth, Alan H., *The Inflationary Universe – The Quest for a New Theory of Cosmic Origins*, Perseus Books, Reading, Massachusetts, 1997.

Han, M Y, *Quarks and Gluons – A Century of Particle Charges*, World Scientific, Singapore, 1999.

Hawking, Stephen, *Black Holes and Baby Universes, and Other Essays*, Bantam Books, New York, paperback Edition, 1994.

Hawking, Stephen & Leonard Mlodinow, *The Grand Design*, Bantam Books, New York, 2010.

Hawking, Stephen, *A Brief History of Time - From the Big Bang to Black Holes*, Bantam Books, New York, 1990.

Hawking, Stephen, *The Theory of Everything – The Origin and Fate of the Universe*, Jaico Publishing House, 2006.

Impey, Chris, *Einstein's Monsters – The Life and Times of Blackholes*, W.W. Norton & Co., New York, 2019.

Impey, Chris, *How It Began – A Time Traveler's Guide to the Universe*, W.W. Norton & Co., New York, paperback Edition, 2013.

Hossenfelder, Sabine, *Lost in Math – How Beauty Leads Physics Astray*, Basic Books, Paperback Edition, New York, NY, 2020.

Hossenfelder, Sabine, *Existential Physics – A Scientist's Guide to Life's Biggest Questions,* Viking, New York, NY, 2022.

Huang, Kerson, *Fundamental Forces of Nature -- The Story of Gauge Fields*, World Scientific Publishing Co., Singapore, 2007.

James, Tim, *Astronomical -From Quarks to Quasars, the Science of Space at Its Strangest*, Robinson/ Little, Brown Book Group, London, 2020.

James, Tim, *Fundamental – How Quantum and Particle Physics Explain Absolutely Everything*, Pegasus Books, New York, 2020.

Jones, Mark, Robert J.A. Lambourne, and Stephen Serjeant, Editors, *An Introduction to Galaxies and Cosmology*, Second Edition, Cambridge University Press, UK, 2015.

Kaku, Michio, *The God Equation – The Quest for a Theory of Everything*, Doubleday, New York, 2021.

Kaku, Michio, *Physics of the Impossible – A Scientific Exploration into the World of Phasers, Force Fields, Teleportation, and Time Travel,* Double Day, New York, 2008.

Kaku, Michio, *Visions- How Science will Revolutionize the 21st Century*, Double Day, New York, 1997.

Kisak, Paul, editor, *The Photon – The Elementary Quantum Particle of Light & Electromagnetic Radiation,* 2022.

Kisslinger, Leonard S., *Astrophysics and The Evolution of the Universe*, World Scientific, Singapore, 2017.

Krauss, Lawrence, *The Greatest Story Ever Told – So Far - Why are We Here*, Simon & Schuster, UK, Paperback Edition, 2018.

Krauss, Lawrence M., *A Universe from Nothing – Why There is Something Rather than Nothing*, Free Press – a Division of Simon & Schuster, New York, 2012.

Krauss, Lawrence, *The Edge of Knowledge -Unsolved Mysteries of the Cosmos*, Post Hill Press, New York, NY, 2023.

Lange, Marc, *An Introduction to The Philosophy of Physics – Locality, Fields, Energy, Mass,* Blackwell Publishing, UK, 2002.

Lincoln, Don, *Einstein's Unfinished Dream*, Oxford University Press, New York, 2023.

Lindley, David, *The Dream Universe -- How Fundamental Physics Lost its Way*, Doubleday, New York, 2020. An easy read recounting the problems with our current theories on how the universe works.

Livio, Mario, *Accelerating Universe – Infinite Expansion, The Cosmological Constant, and the Beauty of the Cosmos*, John Wiley & Sons, Inc., New York and Canada, 2000.

Long, Ariana S., *Ancient Galaxy Clusters Offer Clues About the Early Universe*, Scientific American, Dec. 21, 2021.

Mee, Nicholas, *Gravity*, Oxford University Press, United Kingdom, 2022.

Mersini-Houghton, Laura, *Before the Big Bang – The Origin of the Universe and What Lies Beyond,* Mariner Books, Harper Collins Publishers, New York, 2022.

Munowitz, Michael, *Knowing -- The Nature of Physical Law*, Oxford University Press, New York, 2005.

Musser, George, *Spooky Action at a Distance – The Phenomenon that Reimagines Space and Time and What it Means for Black Holes, the Big Bang, and Theories of Everything*, Scientific American/Farrer, Strauss and Giroux, 2015.

Natarajan, Priyamvada, *Mapping the Heavens,* Yale University Press, New Haven, 2016.

Nomura, Yasunori & Bill Poirier & John Terning, *Quantum Physics, Mini Black Holes, and the Multiverse – Debunking Common Misconceptions in Theoretical Physics, Multiversal Journeys*, Springer International, Switzerland, 2018.

Padilla, Antonio, *Fantastic Numbers and Where to Find Them – A Cosmic Quest from Zero to Infinity,* Farrar, Straus and Giroux, New York, 2022.

Perlov, Delia, and Alex Vilenkin, *Cosmology for the Curious*, Springer International Publishing AG, 2017.

Philips, Raymond, G., *Truth and Absurdity in Modern Physics*, 2021.

Randall, Lisa, *Dark Matter and the Dinosaurs – The Astonishing Interconnectedness of the Universe*, Harper Collins Publishers, New York, 2015.

Randall, Lisa, *Knocking on Heaven's Door – How Physics and Scientific Thinking Illuminate the Universe and the Modern World*, Harper Collins Publishers, New York, paperback edition, 2012.

Rezzolla, Luciano, *The Irresistible Attraction to Gravity – A Journey to Discover Black Holes,* Cambridge University Press, UK, 2023.

Rothman, Tony, *A Little Book About the Big Bang*, Harvard University Press, Cambridge, Mass., 2022.

Rovelli, Carlo, *Reality is Not What it Seems -- The Journey to Quantum Gravity*, Riverhead Books, New York, 2017.

Schmitz, Wouter, *Particles, Fields, and Forces,* Springer Nature Switzerland, Switzerland, 2019.

Schumm, Bruce A., *Deep Down Things -- The Breath-Taking Beauty of Particle Physics*, John Hopkins University Press, Baltimore, 2004.

Seife, Charles, *Alpha & Omega – The Search for the Beginning and End of the Universe*, Viking/Penguin Group, New York, 2003.

Siegel, Ethan, *What Rules the Proton: Quarks or Gluons?,* Forbes Magazine, 2021.

Smethurst, Becky, *A Brief History of Blackholes – And Why Nearly Everything You Know About Them is Wrong,* Macmillan Publishing, London, 2022.

Smith, Timothy Paul, *Hidden Worlds – Hunting for Quarks in Ordinary Matter*, Princeton University Press, Princeton and Oxford, 2003.

Smolin, Lee, *Einstein's Unfinished Revolution – The Search for What Lies Beyond the Quantum*, Penguin Press, New York, 2019.

Smolin, Lee, *The Trouble with Physics – The Rise of String Theory, the Fall of Science, and What Comes Next*, Houton Miffin Co., First Mariner Books Edition, New York, 2007.

Sokol, Joshua, *Earliest Black Hole Gives Rare Glimpse of Ancient Universe, Quanta Magazine, Dec. 6, 2017.*

Still, Ben, *Particle Physics Brick by Brick – Atomic and Subatomic Physics Explained in Legos,* Firefly Books, Ltd., New York, 2018.

Sutter, Paul S., *Your Place in the Universe – Understanding Our Big Messy Existence*, Prometheus Books, New York, 2018.

Tyson, Neil deGrasse and Donals Goldsmith, *Origins, -- Fourteen Billion Years of Cosmic Evolution*, W.W. Norton & Co, New York, 2005, re-issued 2014.

Tucker, Wallace H., *Chandra's Cosmos*, Smithsonian Books, Washington D.C., 2017.

Turner, Ben, *Time moved '5 times slower' in the early universe, mind-bending black hole study reveals,* Live Science Magazine, July 2023.

Wilczek, Frank, *A Beautiful Question – Finding Nature's Deep Design*, Penguin Books, 2015.

Wilczek, Frank, *Fundamentals – Ten Keys to Reality*, Penguin Press, New York, 2021.

Index

accretion
  defined, 35
Alexander, Stephon, 115
aligned Tors
  create a magnetic field, 103
alpha decay, 105
Andromeda galaxy, 45
angular momentum, 251
  & orientational inertia, 207
  in larger swirls, 58
  of particles, 124
annihilation
  avoidance, 72
anthropic principle, 191
anti-matter
  integral part of matter, 71
  mystery of, 70
atom
  emergent qualities of, 153
  evolution of, 151
Baggott, Jim, 191
Ball, Philip, 18
Barnes, Luke, 55
Barrow, John, 46, 125
BBT
  and SMBH creation, 42
  observational evidence, 155
  off on the wrong foot, 28
Beckman, Milo, 84
Bell, John, 174
beta decay, 105, 144
big bang
  particle formation, 65
Big Bang Theory
  see BBT, 1

big crunch, 193
binding energy, 123
biological natural selection, 112
blackhole
  affect on time passage, 241
  made from pure energy, 48
  mergers of, 43
blackhole mass
  coupled to stellar mass, 58
Bohm, David, 18, 168
Bohr, Niels, 165, 167
boson
  configuration, 149
Butterworth, Jon, 73, 126
CAGI
  and emergent qualities, 116
  cosmic progress, 152
  defined, 113
  driving evolution, 188
  force that molds Penergy, 196
  its basic tenets, 193
Carroll, Sean, 3, 12, 118, 177, 178
Cascade Effect
  and homogeneity, 118
  description of, 47
  evidence for, 54
  illustration of, 47
Casmir Affect, 85
Cham, Jorge, 75, 126, 128
charge
  defined, 96
  definition of, 141
  imbalance, 145
Clark, Stuart, 1, 226
Clegg, Brian, 2, 237

Cliff, Harry, 1, 3, 24
CMB, 120
collapse of giant gas clouds, 43
color charge, 100, 146
Comb. & Growth Imperative, 113
complexity, 65
composite particles
    & missing anti-matter, 126
    explain intrinsic spin, 128
    first generation, 51
    source of E&M fields, 126
computer simulations
    regarding SMBH creation, 43
Cooper Pair, 148
Copenhagen Interpretation, 167
cosmic evolution
    defined, 112
cosmic history, 185
Cosmic Microwave Background, 156
cosmic puzzle
    foundation of, 78
cosmic recycling, 114
Cosmic Web
    creation of, 59
cosmological constant
    and Penergy medium, 76
Cosmological Principle, 117
coupling capacity, 89
dark energy
    and Penergy medium, 76
dark matter
    and galaxy creation, 59
    and the Cascade Effect, 56
    origin of, 215
Davies, Paul, 113, 216
de Broglie, Louis, 18, 166, 168
de Coulomb, Augustin, 92
deem blue galaxies, 57
delayed choice
    dbl-slit experiment, 172
destiny of the universe, 193
Devereux, Carolyn, iii, 24, 26, 116, 155
diffraction
    particle tunneling, 15

double-slit experiment
    and CMB photons, 157
    consequences of, 181
    delayed choice, 172
    wave-particle duality, 16
down quark
    decay requiring a boson, 105
    possible configuration of, 144
drag on surrounding spacetime, 226
dwarf galaxies
    evidence for Cascade Effect, 55
dynamic gravity
    origin of, 226
early elementary particles
    attributes of, 68
    one of three stable entities, 52
    spin into existence, 49
Einstein
    and entanglement, 173
    and quantum theory, 166
    and singularities, 25
    and the wave function, 178
    as a realist, 167
    field equations, 42
    General Relativity, 76
    photoelectric effect, 166
    time relative to gravity, 239
electric charge
    definition of, 96
electric charge value, 142
electric force
    matter/anti-matter, 92
electromagnetic field
    as a monopole, 98
electromagnetic force, 96
electron
    configuration, 147
electroweak theory, 205
elementary particle
    creation in four stages of, 67
    rules for interaction, 93
elliptical galaxy, 44
EM waves, 130
emergent qualities
    and fantastic experiences, 195

defined, 115
growing more wonderous, 189
of atoms, 153
of Tryks, 128
energy
three categories, 31
two classes of, 29
entanglement assumptions, 174
entropy
defined, 114
equilibrium
for particle stability, 53
Everett, Hugh, 168
evolution
progresses in Levels, 112
exchange of virtual bosons, 158
Faraday, Michael, 12, 92
Field
source of Work Energy, 31
Field Energy
defined, 30
fields
compatibility, 98
field collapse, 17
spin creates forces, 187
wave-like attributes, 14
fine-tuning problem, 191
first relationship to evolve, 78
first-generation particles, 51
flatness problem, 117
fluctuations Penergy medium, 88
force
acting like rubber band, 102
per Standard Model, 84
per Tor Model, 88
forced decay
definition of, 104
Ford, Kenneth, 86, 181
foundation for our universe, 52
Fractional Quantum Hall Effect, 147
Franklin, Ben, 92
Fritzsch, Harald, 125, 205
Fundamental Principles, 28
galactic distribution
evidence for Cascade Effect, 55

Galaxies buzzing in orbits, 57
galaxy
mass, 119
mergers, 43
Geach, James, 14
Gell-Mann Murry, 100
General Relativity
as a TOE, 164
needs tweaking, 56
Gluons
mediator of strong force, 100
Grand Unified Theory, 187
gravitational
energy, 30
field-alternative theory, 222
force, 99
mass, 224
gravitational condensing
& time dilation, 240
gravitational waves, 75
graviton
and gravitational force, 99
force carrying boson, 86
gravity
defined, 99
Gribbin, John, 57, 216
growth in complexity, 114
Guth, Alan, 116
Hawking, Stephen
and anthropic principle, 191
and blackhole theorm, 25
and laws of physics, 2
blackhole contents, 42
description of inflation, 117
final conclusion, 26
new layers of structure, 126
Heisenberg, Werner, 169
Hertog, Thomas, 2, 172
hf=MC², 53
hidden variables, 174
Higgs boson, 204
Higgs field
gives particles mass, 204
Higgs, Peter, 203
high speed compression
& time dilation, 242

homogeneity problem, 116
Hooke, Robert, 166
horizon problem, 116
Hossenfelder, Savine, 197
Hot Big Bang Theory
    creation of nucleus, 150
    regarding SMBH creation, 42
Hubble expansion constant, 155
Hubble Space Telescope, 43
huge voids
    left by swirl disintegration, 50
HUP
    limitation to measuremt, 170
    a math tool, 180
Huygens, Christian, 166
If particles are real, 77
Impey, Chris, 56, 58, 126
inertial mass, 224
inflation
    defined, 116
initial conditions
    posited as, 27
initial spinning speck
    origin angular momentum, 46
inside a blackhole, 41
intelligence and technology, 198
Intergalactic Penergy Medium
    origin of, 73
intrinsic particle spin, 124
F=Kqq/r², 98
James Webb Space Telescope, 43
James, Tim, 178
Jones, Mark, 181
Journey of the Universe, 194
JWST, 49
    discovery of The Vine, 55
Kaku, Michio, iii, 25, 36, 84, 87, 166
Kinetic Energy
    defined, 31
knot of Penergy
    and particle creation, 206
    creating new particles, 86
Krauss, Lawrence, 87, 196, 204
Large Hadron Collider, 100, 103, 129, 204, 251

first generation particles, 52
large voids
    origin of, 60
largest mismatch in science, 87
Level of matter
    Eighth Level, 194
    Seventh Level of matter, 190
    Six Levels listed, 190
    Third Level, 112
levels of evolution described, 112
Lewis, Geraint, 55
Lincoln, Don, 125, 199, 205
Lindley, David, 125, 205
Livio, Mario, 74
locality, 174
Logical Positivism, 167
magnetic field
    created by Tor-Chain, 91
magnetic force
    per Tor Model, 88
magnetic monopoles, 98
magnetism and electricity
    relationship between, 135
major sci. accomplishments, 33
majorana
    defined, 136
Many-Worlds Theory, 168
mass
    an alternative theory for, 207
Mass Energy and Work Energy
    difference between, 33
mass of a blackhole
    more about Penergy, 42
mathematics
    essential for survival, 197
matter and anti-matter, 69
Maxwell, 33
measurement problem, 176
Mersini-Houghton, Laura, 75
meson, 146
Milgrom, Mordehai, 230
minute environment
    of a particle, 181
mirror image, 69
Modified Newtonian Dynamics, 230

monopoles, 99, 117
Munowitz, Michael, 12, 105, 126
Musser, George, 12, 86
natural decay, 104
natural selection, 152
    and particle creation, 127
    between particles, 111
network between particles, 12
neutrino
    as a connector, 136
    oscillation, 137
neutron
    decay, 103
    evolutionary purpose, 151
Newton, Issac, 166
nexus
    between mass & gravity, 224
NIB-state
    crushing Penergy to, 227
    definition of, 27
    within a blackhole, 42
Nomura, Yasunori, 177, 181
now
    no such thing as, 171
nuclear force, 151
nucleus evolution
    per Standard Model, 149
    per Tor Model, 150
observational evidence
    for SMBH creation, 54
orientational inertia, 208
particle
    angular momentum, 124
    annihilation, 69
    behaving like wave, 170
    decay, 104
    entanglement, 173
    never seen as a wave, 13
    pairs-creation of, 67
    resilence, 69
    spin creates fields, 187
    spin signature, 72
    two turn mystery, 124
    virtual, 74, 85
particle evolution
    sesquence of, 154

particle field collapse, 17
particle fields
    in Tor Model, 13
particle tunneling, 15
particle wave frequency
    equal to spin frequency, 53
particles
    combine in two ways, 123
    touching, 97
Pauli Exclusion Principle, 147
Penergy
    all consumed in, 52
    defined, 29
    is a physical energy, 77
    natural attributes, 34
    NIB state, 27
    same as mass energy, 32
Penergy density
    controls Penergy behavior, 36
    related to Relativity, 239
Penergy medium
    cosmic communication, 175
    the definition of, 76
Penrose, Roger, 25
personal field
    description of, 13
    double-slit experiment, 16
perspective
    on life, 190
    The CAGI, 193
    The Cosmic, 194
photoelectric effect, 166
photon
    absorption, 134
    configuration, 131
    decay, 133
    energy-definition of, 30
    not an EM wave, 130
    polarization-entanglement,
        173
    spin rate, 132
    travels in straight line, 237
physical energy
    defined, 29
pilot wave theory, 18, 168
Pion, 146

Planck's equation, E=hf, 53
Prediction
    creation of Penergy dynamics,
        188
    extrapolate Bohm's math, 18
    force through particle fields,
        107
    loss of attractive force, 97
    mixing matter & anti-matter,
        72
    new origin for mass, 212
    origin gravitational field, 225
    SMBH creation, 61
    SMBH gravitational field, 229
Preons, 73
proton
    investigation of interior, 101
proto-stars, 62
pure energy
    blackholes, 48
    interspatial medium, 76
    Penergy, 23
QFT
    particle field creation, 11
QT is a ridiculous theory, 166
Quantum
    Electrodynamics, 85
    Gravity, 2
    weirdness, 170
quantum mechanics
    accuracy attributed to, 19
    Dbl-slit prediction, 171
    why so accurate, 177
quantum potential, 18
quantum theory
    and entanglement, 174
    and wave function, 176
    wave-particle duality, 15
quark
    complex configuration, 144
    decay, 144
    formation charge values, 142
    simply configuration, 143
Randall, Lisa, 75, 186
reality
    of evolution, 189

of Quantum Theory, 177
reductionist philosophy, 195
Rees, Martin, 43
relationships
    power things to move, 36
renormalization
    mathematically dubious, 85
Rezzolla, Luciano, 199, 228
Rothman, Tony, 118
schematic mechanical universe,
    33
Schmitz, Wouter, 124
Schrodinger equation, 176
Schumm, Bruce, 73, 125
Schwarzschild, Karl, 25, 42, 227
second law thermodynamics, 114
second relationship to evolve, 78
sequence of particle evol., 154
Siegel, Ethan, 101
Silk, Joseph, 125
singularity
    inside blackhole, 42
Singularity Problem, 25
SMBH
    evidence supporting creation,
        54
    less than 50K solar masses, 40
SMBH creation
    scientists hamstrung, 42
Smith, Timothy Paul, 211
Smolin, Lee, 164, 170, 181
Space
    description of, 74
spacetime
    why relative, 235
Special Relativity, 33
spin
    creates tension in medium, 35
    density relationship, 49
    dynamics of, 89
    importance of, 69
    natural trait of Penergy, 35
spin-rate to mass ratio, 51
spin-space, 125
spiral galaxies
    mergers of, 44

spooky action at a distance, 173
Standard Model
 and forces, 83
 and the weak force, 103
stars
 formation of, 46
 origin of first, 62
stellar collapse, 41
Still, Ben, 204, 211
Strange Quark and Muon, 51
String Theory, 186
strong force
 per Standard Model, 100
Super TOE, 188
supermassive blackholes
 See SMBH, 54
superposition
 is dubious, 180
Supersymmetry, 186
support for composites, 126
Sutter, Paul, 25, 56, 74
swirl disintegration, 50
theoretical energy value, 87
Theory of Everything, 186
time dilation, 240
 bouncing light scenario, 237
time=0, 27
Top Quarks and Tau Leptons, 51
Tor Model
 definition of, 66
 particle location, 178
 quarks made from Tryks, 141
 wave function, 176
Tor spin
 and gyroscopic effect, 207
 and magnetic force, 89
 and strong force, 102
Tor-Chain
 bundle, 94
 creation of, 91
Torons, 66
Tors
 emergent qualties of, 116
 not subject to annihilation, 95

opposite spinning, 96
spin dynamics, 102
spin on their own axis, 68
Tryk-Chain wobble, 135
Tryks, 111
 creation of EM field, 128
 emergent qualities of, 128
 illustration of, 96
Tucker, Wallace H., 59
Turner, Ben, 243
ubiquitous fields, 11
uncertainty
 examination of, 168
universal constants, 191
universe
 charge neutral, 70
 discrepancy in age, 155
 surely on a journey, 195
vacuum of space
 and requisite energy, 149
 awash in virtual particles, 87
 energy borrowed, 74
 short on energy, 104
virtual particles
 existence of, 74
 in Tor Model, 87
virtual photons, 84
W and Z bosons, 104
wave particle duality
 may not be real, 179
 Pilot Wave Theory, 18
wavefunction
 DBL-Slit experiment, 171
 examination of, 175
 quantum mechanics, 167
weak force
 theoretical development, 103
web-like appearance, 61
Wheeler, John, 172
Whiteson, Daniel, 75, 126, 128
Wilczek, Frank, 147, 165, 207
work energy
 defined, 30
 derived from, 32

About the Author: Robert J. Conover

After earning a degree in philosophy, Bob took up studying particle physics and cosmology with a passion. Intrigued by the new discoveries taking place in particle physics and their influence on our changing picture of the early universe, he was drawn into finding answers to the many questions, conflicts, and unknowns that arose from those discoveries.

A New Vision of the Early Universe is a culmination of forty years of research and study of the writings of particle physicists, cosmologists, and mathematicians. The depth of Bob's research, attention to detail, clear prose, and philosophical approach to problem solving conspire to make this a well-respected text. Unrelentless in his pursuit of a viable vision of the early universe, Bob has amassed additional material supporting a broader perspective on how the universe evolved, necessitating this latest Edition.

Bob has retired from a career as a civil investigator and mediation negotiator, during which time he authored *Strategic Tort-Mediation Negotiation*. After retirement he penned *Journey of the Universe*, a compelling theory as to what is driving cosmic evolution.

Bob resides with his wife Ellen in San Luis Obispo, California, where he continues to research, write, create board games, and cultivate the minds of their eight grandchildren.